Advanced
Modular
Mathematics

Pure
Mathematics 3

Stephen Webb

**SECOND
EDITION**

Unit P3

Published by HarperCollins Publishers Limited
77–85 Fulham Palace Road
Hammersmith
London W6 8JB

> www.CollinsEducation.com
> On-line Support for Schools and Colleges

© National Extension College Trust Ltd 2000
First published 2000
ISBN 000 322511 9

This book was written by Stephen Webb for the National Extension College Trust Ltd.

All rights reserved. No part of this publication may be reproduced, stored in a retrieval system, or transmitted in any form or by any means, electronic, mechanical, photocopying, recording or otherwise, without either the prior permission of the Publisher or a licence permitting restricted copying in the United Kingdom issued by the Copyright Licensing Agency Ltd, 90 Tottenham Court Road, London W1P 0LP.

British Library Cataloguing in Publication Data
A catalogue record for this publication is available from the British Library.

Original internal design: Derek Lee
Cover design and implementation: Terry Bambrook
Project editors: Hugh Hillyard-Parker and Margaret Levin
Page layout: Mary Bishop
Printed and bound: Scotprint, Musselburgh

The authors and publishers thank Dave Wilkins for his comments on this book.

The National Extension College is an educational trust and a registered charity with a distinguished body of trustees. It is an independent, self-financing organisation.

Since it was established in 1963, NEC has pioneered the development of flexible learning for adults. NEC is actively developing innovative materials and systems for distance-learning options from basic skills and general education to degree and professional training.

For further details of NEC resources that support Advanced Modular Mathematics, and other NEC courses, contact NEC Customer Services:

National Extension College Trust Ltd
18 Brooklands Avenue
Cambridge CB2 2HN
Telephone 01223 316644, Fax 01223 313586
Email resources@nec.ac.uk, Home page www.nec.ac.uk

You might also like to visit:

> www.fireandwater.com
> The book lover's website

UNIT P3

Contents

Section 1 Algebra — 1
Partial fractions — 1
Quadratic factors — 2
Repeated factors — 4
Top-heavy fractions — 5
The remainder theorem — 7

Section 2 Coordinate geometry — 13
The equation of a circle — 13
Tangents to a circle — 14
Points of intersection — 16
Parametric equations — 17
Finding the cartesian equation — 17
Curve sketching — 21
Transformations of the curve $y = \dfrac{1}{x}$ — 25

Section 3 Series — 31
General binomial expansion — 31
Multiple series — 33
Rearrangements — 34
Larger terms — 35
Use of partial fractions — 36
Use in approximation — 37

Section 4 Differentiation — 41
Trigonometric functions — 41
Differentiation of a product — 41
Differentiation of a quotient — 42
Implicit differentiation — 43
Differentiation of a^x — 44
A function of a function — 45
Powers of a function — 48
Derivatives of $\sec x$, $\operatorname{cosec} x$, $\cot x$ — 49
Partial fractions and differentiation — 50
Functions of $\ln x$ — 51
Parametric differentiation — 52
Tangents and normals — 54

Section 5 Integration — 61
Standard integrals — 61
Recognition — 62
Powers of linear functions — 64
Integrals with partial fractions — 64
Substitution — 66
Changing the variable — 66
Changing the limits — 69
Integration by parts — 70
Integration by parts – twice — 72
Trigonometric integration — 74
Using identities — 75
Areas — 77

Section 6 First-order differential equations — 85
Separable variables — 85
Solving the equations — 86
Partial fractions — 89
Rates of change — 91
Related rates of change — 96
More than one factor — 98

Section 7 Vectors — 105
Scalars and vectors — 105
Adding displacement vectors — 106
Multiplication by a scalar — 109
The vector equation of a line — 110
Equation of a line passing through two given points — 113
Vectors in three dimensions — 114
Special cases — 116
Intersection of two lines — 117
The length of vectors — 119
Unit vectors — 120
The scalar product — 121
The angle between vectors — 122
Angles between lines — 124
Perpendicular vectors — 125

Practice examination paper — 135

Solutions to exercises — 137

UNIT P3

Advanced Modular Mathematics

FOREWORD   This book is one of a series covering the Edexcel Advanced Subsidiary (AS) and Advanced GCE in Mathematics. It covers all the subject material for Pure Mathematics 3 (Unit P3), examined from 2001 onwards.

While this series of text books has been structured to match the Edexcel specification, we hope that the informal style of the text and approach to important concepts will encourage other readers whose final exams are from other Boards to use the books for extra reading and practice. In particular, we have included references to the OCR syllabus (see below).

This book is meant to be *used*: read the text, study the worked examples and work through the Practice questions and Summary exercises, which will give you practice in the basic skills you need for maths at this level. Many exercises, and worked examples, are based on applications of the mathematics in this book. There are many books for advanced mathematics, which include many more exercises: use this book to direct your studies, making use of as many other resources as you can.

There are many features in this book that you will find particularly useful:

- Each **section** covers one discrete area of the new Edexcel specification. The order of topics is exactly the same as in the specification.

- **Practice questions** are given at regular intervals throughout each section. The questions are graded to help you build up your mathematical skills gradually through the section. The **Answers** to these questions come at the end of the relevant section.

- **Summary exercises** are given at the end of each section; these include more full-blown, exam-type questions. Full, worked solutions are given in a separate **Solutions** section at the end of the book.

- In addition, we have provided a complete **Practice examination paper**, which you can use as a 'dummy run' of the actual exam when you reach the end of your studies on P3.

- Alongside most of the headings in this book you will see boxed references, such as: OCR **P3** 5.3.1 (a) These are for students following the OCR specification and indicate which part of that specification the topic covers.

- Key Skills: because of the nature of pure mathematics, your work on this book will not provide many obvious opportunities for gathering evidence of Key Skills, and so we have not included any Key Skills references (as we have done in other books in this series).

The National Extension College has more experience of flexible-learning materials than any other body (see p. ii). This series is a distillation of that experience: Advanced Modular Mathematics helps to put you in control of your own learning.

Permissions

We are grateful to the following examination boards for permission to reproduce in the Summary exercises and Practice questions the following questions from past examination papers.

The examination boards credited below accept no responsibility whatsoever for the accuracy or method of working in the answers given in the Solutions section, which are entirely the responsibility of the author. Please note that the solutions given may not necessarily constitute the only possible solutions.

Edexcel
(including questions published by the London Board)

Section 2:   Exercises 9, 10
Section 3:   Exercises 2, 5–7, 9, 11, 12
Section 5:   Practice questions I, nos 2, 3
Section 7:   Exercises 1, 2, 4, 5, 7, 9–12

OCR
(including questions published by the Oxford and Cambridge Boards)

Section 1:   Exercise 1
Section 2:   Exercise 5
Section 3:   Exercises 1, 3, 4, 10
Section 4:   Exercises 1, 2; Practice questions I, nos 6–8
Section 5:   Exercises 1, 2, 4–7; Practice questions F, nos 4, 5
Section 7:   Exercises 3, 6, 8; Practice questions H, no. 7

AQA
(including questions published by the Associated Examining Board)

AQA (AEB) examination questions are reproduced by permission of the Assessment and Qualifications Alliance.

Section 2:   Exercises 1–4, 6
Section 3:   Exercise 8
Section 7:   Practice questions I, no. 6

SECTION 1

Algebra

INTRODUCTION In this section we are going to see how we can split certain types of rational functions into their component parts. This makes certain operations that we look at later in this unit easier to carry out: rewriting as a series, integration and differentiation. We shall also extend the idea of the factor theorem from P1 to see how we can find remainders without having to carry out long algebraic division.

Partial fractions

OCR P3 5.3.1 (b)

We have seen how we can combine two or more fractions into a single rational function – we're now going to work the other way round. Instead of learning how to combine fractions, we're going to see how we can split them into their components.

For example, we can combine the fractions $\frac{2}{x-1}$ and $\frac{3}{x-2}$:

$$\frac{2}{x-1} + \frac{3}{x-2} = \frac{(x-2) \times 2 + (x-1) \times 3}{(x-1)(x-2)}$$
$$= \frac{2x - 4 + 3x - 3}{(x-1)(x-2)}$$
$$= \frac{5x - 7}{(x-1)(x-2)}$$

Working the other way round, if we have the fraction:

$$\frac{x+7}{(x-2)(x+1)}$$

can we find the two numbers A and B such that:

$$\frac{x+7}{(x-2)(x+1)} = \frac{A}{(x-2)} + \frac{B}{(x+1)}?$$

The method of finding these numbers is to put the two fractions together:

$$\frac{A}{(x-2)} + \frac{B}{(x+1)} = \frac{A(x+1) + B(x-2)}{(x-2)(x+1)}$$

and compare this with what we want, i.e.

$$\frac{A(x+1) + B(x-2)}{(x-2)(x+1)} = \frac{x+7}{(x-2)(x+1)}$$

We can see that the bottom lines are the same – we have to choose A and B so that the tops are also the same, i.e. so that:

$$A(x+1) + B(x-2) = x+7$$

This is to be true for any value of x, i.e. it is an *identity* and we can find the unknown constants by the methods we have already looked at earlier in P2. Let's rewrite it with the extra line in the equals sign:

$$A(x+1) + B(x-2) \equiv x+7$$

Since this is to be true for *any* value of x, we can substitute values that suit us. So we choose values that make the value of the brackets zero in turn, i.e.

putting $x = -1$: $A(-1+1) + B(-1-2) = -1+7$
$$-3B = 6$$
$$B = -2$$

putting $x = 2$: $\quad 3A = 9 \Rightarrow A = 3$

So our original fraction:

$$\frac{x+7}{(x-2)(x+1)} \equiv \frac{3}{(x-2)} - \frac{2}{(x+1)}$$

which we can verify by recombining the fractions on the right-hand side. These two fractions are said to be the *partial fractions* of the original fraction.

Let's have a look at another example of this.

Example

Given that $f(x) \equiv \dfrac{1}{(1+x)(1-2x)}$

express $f(x)$ in partial fractions.

Solution

Let $\quad \dfrac{1}{(1+x)(1-2x)} \equiv \dfrac{A}{(1+x)} + \dfrac{B}{(1-2x)}$

$$\equiv \frac{A(1-2x) + B(1+x)}{(1+x)(1-2x)}$$

Then Putting $x = \tfrac{1}{2}$: $\quad 1 = \left(\tfrac{3}{2}\right)B \Rightarrow B = \tfrac{2}{3}$

Putting $x = -1$: $\quad 1 = 3A \Rightarrow A = \tfrac{1}{3}$

i.e. $f(x) \equiv \dfrac{1}{3(1+x)} + \dfrac{2}{3(1-2x)}$

Practice questions A

1. Put the following into partial fractions

(a) $\dfrac{4x+1}{(x+1)(2x-1)}$ (b) $\dfrac{2x+1}{(x+2)(x+1)}$ (c) $\dfrac{x-2}{(x+4)(x+1)}$ (d) $\dfrac{4}{(x-3)(x+1)}$

(e) $\dfrac{4}{x^2-4}$ (f) $\dfrac{x+8}{2x^2-3x-2}$ (g) $\dfrac{2}{x(x+1)(x-1)}$ (h) $\dfrac{2-x}{x^2-1}$

② Quadratic factors

OCR P3 5.3.1 (b)

The next step is to look at the case when one of the factors in the denominator is quadratic, which means that the highest power of x occurring is two, something like $2x^2 + 3$ for example.

It could be that the quadratic factor can be factorised, e.g. $x^2 - 4$ could be expressed as $(x - 2)(x + 2)$, in which case all the factors are linear and the method above can be used. Otherwise, if the quadratic factor is *irreducible*, i.e. it cannot be put into real factors, we need to modify the method and use a different form of numerator (i.e. top of the fraction).

Let's take an example to see why this is so. Suppose we wanted to put the following function into partial fractions:

$$f(x) \equiv \frac{3x^2 + 3x + 2}{(x + 1)(x^2 + 1)}$$

If we suppose the two fractions are:

$$\frac{A}{x + 1} + \frac{B}{x^2 + 1}$$

we find that we need to choose A and B so that:

$$3x^2 + 3x + 2 \equiv A(x^2 + 1) + B(x + 1)$$

x^2-coeff: $3 = A$

x-coeff: $3 = B$

constant: $2 = A + B$

This last equation is not consistent with the other two, so we could never find suitable values for A and B. Instead, we suppose that the numerator of the quadratic factor is of the form $Bx + C$, i.e. put:

$$\frac{3x^2 + 3x + 2}{(x + 1)(x^2 + 1)} \equiv \frac{A}{x + 1} + \frac{Bx + C}{x^2 + 1}$$

$$\equiv \frac{A(x^2 + 1) + (Bx + C)(x + 1)}{(x + 1)(x^2 + 1)}$$

Then $3x^2 + 3x + 2 = A(x^2 + 1) + (Bx + C)(x + 1)$

Put $x = -1$: $3 - 3 + 2 = 2A$

$\Rightarrow$ $A = 1$

There are no values of x that will make the bracket $(x^2 + 1)$ disappear, so to find B and C we use a different approach. Let's multiply the right-hand side out completely:

$$3x^2 + 3x + 2 \equiv Ax^2 + A + Bx^2 + Bx + Cx + C$$
$$\equiv (A + B)x^2 + (B + C)x + (A + C)$$

when we arrange the terms in descending powers of x. Now the fact that this is an identity means that *each* of the coefficients must be *identical on both sides*, i.e.:

x^2-coeff $3 = A + B$

x-coeff $3 = B + C$

constant $2 = A + C$

With the value of A we've already found, we can work out the other values:

$A = 1, B = 2, C = 1$

i.e. $\dfrac{3x^2 + 3x + 2}{(x + 1)(x^2 + 1)} \equiv \dfrac{1}{x + 1} + \dfrac{2x + 1}{x^2 + 1}$

In an actual question, we wouldn't write down the whole of the expansion of the right-hand side. As we did before, we would find the constant belonging with the linear factor (A in the last example) and then mentally find the coefficient of the x^2 term and the constant – these are usually the most straightforward to find. Let's work through an example of this.

Example

Express $f(x) = \dfrac{x+6}{(x+1)(x^2+4)}$ in partial fractions.

Solution

Let $\dfrac{x+6}{(x+1)(x^2+4)} \equiv \dfrac{A}{x+1} + \dfrac{Bx+C}{x^2+4}$

$\equiv \dfrac{A(x^2+4) + (Bx+C)(x+1)}{(x+1)(x^2+4)}$

i.e. $x+6 \equiv A(x^2+4) + (Bx+C)(x+1)$ ... ①

Put $x = -1$: $5 = 5A \Rightarrow A = 1$

(Now look on the right hand side of ① for the coefficients of x^2; without multiplying out completely you can see that this is $A + B$. Similarly the constant must be $4A + C$. If you can't see this directly, then multiply out and collect together as we did in the previous example.)

Comparing coefficients of x^2 :

$0 = A + B \Rightarrow B = -1$

(The coefficient of x^2 on the left-hand side is 0 because there is no x^2 term.)

Comparing constants:

$6 = 4A + C \Rightarrow C = 2$

Then $\dfrac{x+6}{(x+1)(x^2+4)} \equiv \dfrac{1}{x+1} + \dfrac{2-x}{x^2+4}$

Practice questions B

1. Put the following into partial fractions

(a) $\dfrac{5x}{(x+2)(x^2+1)}$ (b) $\dfrac{2x^2-6x-7}{(2x+3)(x^2+1)}$ (c) $\dfrac{3x+1}{(x+1)(x^2+1)}$ (d) $\dfrac{3x+1}{x(x^2+1)}$

(e) $\dfrac{x+1}{x(x^2+x+1)}$ (f) $\dfrac{3x+5}{(x-2)(2x^2+3)}$ (g) $\dfrac{8x-1}{(x-2)(x^2+1)}$ (h) $\dfrac{x+2}{(x-2)(x^2+4)}$

③ Repeated factors

OCR P3 5.3.1 (b)

There is another type of fraction to look at: one of the factors on the bottom can be a 'repeated factor', something like $(x-2)^2$. Since this is a quadratic, we would expect to put $Ax + B$ on top as we did with the type above; instead we have the slightly curious procedure of putting:

$\dfrac{A}{x-2} + \dfrac{B}{(x-2)^2}$

This makes it easier to deal with in later work.

Let's have a look at an example of this type. You'll notice that the gathering together of the *three* fractions is a little more complicated.

Algebra P3

Example Express $\dfrac{11 - x - x^2}{(x + 2)(x - 1)^2}$ in partial fractions.

Solution Let $\dfrac{11 - x - x^2}{(x + 2)(x - 1)^2} \equiv \dfrac{A}{x + 2} + \dfrac{B}{x - 1} + \dfrac{C}{(x - 1)^2}$

$$\equiv \dfrac{A(x - 1)^2 + B(x + 2)(x - 1) + C(x + 2)}{(x + 2)(x - 1)^2}$$

You multiply the top by what isn't on the bottom, but *is* in the common denominator, i.e.:

$$11 - x - x^2 \equiv A(x - 1)^2 + B(x + 2)(x - 1) + C(x + 2)$$

At least with this type there are two values of x which make brackets disappear:

putting $x = 1$: $\quad 9 = 3C \quad \Rightarrow \quad C = 3$

putting $x = -2$: $\quad 9 = 9A \quad \Rightarrow \quad A = 1$

x^2-coeffs.: $\quad -1 = A + B \quad \Rightarrow \quad B = -2$

i.e. $\dfrac{11 - x - x^2}{(x + 2)(x - 1)^2} \equiv \dfrac{1}{x + 2} - \dfrac{2}{x - 1} + \dfrac{3}{(x - 1)^2}$

Practice questions C

1 Put the following into partial fractions

(a) $\dfrac{4 + 7x}{(2 - x)(1 + x)^2}$ (b) $\dfrac{2x^2 + 1}{x(x - 1)^2}$ (c) $\dfrac{x^2 - 11}{(x + 2)^2(3x - 1)}$ (d) $\dfrac{x - 4}{(2x + 1)(x - 1)^2}$

(e) $\dfrac{1}{x^2(x - 1)}$ (f) $\dfrac{5x + 7}{(x + 1)^2(x + 2)}$ (g) $\dfrac{x}{(x - 1)^2}$ (h) $\dfrac{3x^2 + 2}{x(x - 1)^2}$

④ Top-heavy fractions   OCR P4 5.4.1 (a)

The final case is when the greatest power of x of the top of the fraction (i.e. the degree of the numerator) is greater than or equal to the degree of the bottom (the denominator).

When the degree of the numerator is the *same* as the degree of the denominator, the partial fractions will be of the form

$$A + f(x)$$

where $f(x)$ is the standard expression for this type of partial fraction. We can find A quite simply by inspection: for example

$$\dfrac{x^2}{(x - 1)(x - 2)} \equiv 1 + \dfrac{B}{x - 1} + \dfrac{C}{x - 2}$$

since x^2 (on the top) divided by x^2 (on the bottom) gives $A = 1$.

Similarly, $\dfrac{3x^3}{(2x - 1)(x^2 + 4)} \equiv \dfrac{3}{2} + \dfrac{B}{2x - 1} + \dfrac{Cx + D}{x^2 + 4}$

since $3x^3$ (on the top) divided by $2x^3$ (from the bottom) gives $A = \dfrac{3}{2}$

Example

Express $\dfrac{x^2}{(x-1)(x-2)}$

Solution

We have already seen that we can put this as

$$\dfrac{x^2}{(x-1)(x-2)} \equiv 1 + \dfrac{B}{x-1} + \dfrac{C}{x-2}$$

The fractions on the RHS recombine to give

$$\dfrac{(x-1)(x-2) + B(x-2) + C(x-1)}{(x-1)(x-2)}$$

and so our identity is

$$x^2 \equiv (x-1)(x-2) + B(x-2) + C(x-1)$$

$x = 1 \quad 1 = B(-1) \Rightarrow B = -1$
$x = 2 \quad 4 = C \quad\quad \Rightarrow C = 4$

i.e. $\dfrac{x^2}{(x-1)(x-2)} = 1 - \dfrac{1}{x-1} + \dfrac{4}{x-2}$

Example

Express $\dfrac{2x^3 - 2x^2 - x - 1}{(x-2)(x^2+1)}$ in partial fractions.

Solution

We see that the required form is

$$2 + \dfrac{B}{x-2} + \dfrac{Cx + D}{x^2 + 1}$$

This recombines to give $\dfrac{2(x-2)(x^2+1) + B(x^2+1) + (x-2)(Cx+D)}{(x-2)(x^2+1)}$

with identity

$$2x^3 - 2x^2 - x - 1 \equiv 2(x-2)(x^2+1) + B(x^2+1) + (x-2)(Cx+D)$$

$x = 2$: $\quad 16 - 8 - 2 - 1 \quad = 5B \Rightarrow B = 1$
constant: $\quad\quad\quad -1 \quad\quad = -4 + B - 2D \Rightarrow D = -1$
x-coeff: $\quad\quad\quad -1 \quad\quad = 2 - 2C + D \Rightarrow C = 1$

i.e. $\dfrac{2x^3 - 2x^2 - x - 1}{(x-2)(x^2+1)} \equiv 2 + \dfrac{1}{x-2} + \dfrac{x-1}{x^2+1}$

When the degree of the numerator is greater than that of the denominator and we are not given any lead-in, we may have to use algebraic division.

Example

Express $\dfrac{x^3}{(x+2)(x-1)}$ in partial fractions.

Solution

Multiplying out, the denominator is $x^2 + x - 2$ and so the division is

$$\begin{array}{r}
x - 1 \\
x^2 + x - 2 \overline{\smash{\big)}\, x^3 } \\
\underline{x^3 + x^2 - 2x} \\
-x^2 + 2x \\
\underline{-x^2 - x + 2} \\
3x - 2
\end{array}$$

i.e. $\dfrac{x^3}{(x+2)(x-1)} \equiv x - 1 + \dfrac{3x-2}{(x+2)(x-1)}$

Using the standard method,

$$\dfrac{3x-2}{(x+2)(x-1)} \equiv \dfrac{A}{x+2} + \dfrac{B}{x-1}$$

$\Rightarrow \quad 3x - 2 \equiv A(x-1) + B(x+2)$

$x = 1 \qquad 1 = 3B \Rightarrow B = \dfrac{1}{3}$

$x = -2 \qquad -8 = -3A \Rightarrow A = \dfrac{8}{3}$

i.e. $\dfrac{x^3}{(x+2)(x-1)} \equiv x - 1 + \dfrac{8}{3(x+2)} + \dfrac{1}{3(x-1)}$

Practice questions D

1 Express the following in partial fractions

(a) $\dfrac{x^2}{(x+2)(x+3)}$ (b) $\dfrac{x^2+1}{(x-1)(x+1)}$ (c) $\dfrac{x^2+x+1}{(x-1)(x+1)}$ (d) $\dfrac{x^3+1}{x(x-1)}$

(e) $\dfrac{x^3}{(x-1)(x+1)}$ (f) $\dfrac{x^2-x-5}{x^2-x-6}$ (g) $\dfrac{2x^3+x^2-15x-5}{(x+3)(x-2)}$ (h) $\dfrac{2x^3-x-1}{(x-3)(x^2+1)}$

2 Find the constants A, B and C such that

$$\dfrac{x^2-5}{x-2} \equiv Ax + B + \dfrac{C}{x-2}$$

We are now going to extend out work on polynomials that was begun in P1.

The remainder theorem

OCR P2 5.2.3 (e)

Suppose we had the polynomial:

$P(x) = x^3 - 3x^2 + 5x - 3$

If we divided this by $(x-2)$, what remainder would we be left with? We could find out by long division:

```
                x² - x + 3
       ┌─────────────────────
x - 2  │ x³ - 3x² + 5x - 3
         x³ - 2x²
         ─────────
              -x² + 5x
              -x² + 2x
              ─────────
                    3x - 3
                    3x - 6
                    ───────
                         3
```

So the remainder is 3, with a quotient of $x^2 - x + 3$. Have a go yourself, dividing the same polynomial $P(x)$ by $(x + 2)$.

You should have ended up with a remainder of –33 and a quotient of $x^2 - 5x + 15$. (If not, check that you have taken each line away correctly from the one above.)

This method is quite time-consuming and, as you may have found out, prone to producing algebraic slips.

There is a quicker way of finding the remainder. If we do not need to know the quotient we use the *remainder theorem*.

> **Remainder theorem**
>
> If a polynomial P(x) is divided by ($x - a$)
>
> then the remainder is P(a)

Let's check this with the remainders we have already found:

When we divide P(x) by ($x - 2$), $a = 2$ and
$$P(2) = 8 - 3 \times 4 + 5 \times 2 - 3 = 3$$

With the one you tried, ($x + 2$), $a = -2$ and
$$P(-2) = -8 - 3 \times 4 + 5(-2) - 3 = -33$$

So it works with those two. Can we see why it works? Using the results from our divisions, we could write:
$$P(x) = (x - 2)(x^2 - x + 3) + 3$$
and $P(x) = (x + 2)(x^2 - 5x + 15) - 33$

In fact, if we divide P(x) by ($x - a$), where a is any number, we will get a quotient, which we can call Q(x), and a remainder R. We could write:
$$P(x) \equiv (x - a)Q(x) + R$$

If we substitute $x = a$ on both sides of this identity, the bracket ($x - a$) is zero and we have:
$$P(a) = 0 \times Q(a) + R = R$$

and this is what the remainder theorem states.

Example   What is the remainder when $x^4 - 3x^3 + x - 6$ is divided by

(a) ($x - 3$)   (b) ($x + 2$)?

Solution   Let P(x) $\equiv x^4 - 3x^3 + x - 6$.

(a)   Here $a = 3$, so by the remainder theorem the remainder is:
$$P(3) = 81 - 81 + 3 - 6 = -3$$

(b)   Here $a = -2$, so the remainder is P(-2) = $16 + 24 - 2 - 6 = 32$

Further examples

Here are a couple more examples of different kinds that crop up in exam papers.

Example   If P(x) = $4x^3 - kx^2 + 5x + 8$ leaves a remainder of 2 when divided by ($2x - 3$), find k.

Solution	$P\left(\frac{3}{2}\right) = \frac{27}{2} - \frac{9k}{4} + \frac{15}{2} + 8 = 2$ by remainder theorem. $\Rightarrow 29 - \frac{9k}{4} = 2$, so $k = 12$
Example	The polynomial $2x^3 - 3ax^2 + ax + b$ has a factor of $(x - 1)$ and, when divided by $(x - 2)$, a remainder of -54 is obtained. Find the values of a and b and factorise the polynomial.

Solution	Put $P(x) = 2x^3 - 3ax^2 + ax + b$
	If $(x - 1)$ is a factor, then $P(1) = 0$.
	      $P(1) = 2 - 3a + a + b = 0$
	      $b - 2a = -2$ ... ①
	Also $P(-2) = -54$ by remainder theorem
	      $P(-2) = -16 - 12a - 2a + b = -54$
	      $b - 14a = -38$ ... ②
	Solving ① and ② simultaneously, $a = 3$ and $b = 4$
	Then $P(x) = 2x^3 - 9x^2 + 3x + 4$
	Divide this by the factor $(x - 1)$
	$$\begin{array}{r} 2x^2 - 7x - 4 \\ x-1 \overline{\smash{)}2x^3 - 9x^2 + 3x + 4} \\ \underline{2x^3 - 2x^2} \\ -7x^2 + 3x \\ \underline{-7x^2 + 7x} \\ -4x + 4 \\ \underline{-4x + 4} \end{array}$$
	$P(x) = (x - 1)(2x^2 - 7x - 4)$
	        $= (x - 1)(2x + 1)(x - 4)$
	Note: If the question says factorise completely, the factors will usually all be linear.

Example	Find the remainder and the quotient when $f(x) = x^3 + 2x^2 - 5x + 7$ is divided by $x + 2$.
Solution	Since we need the quotient here, we have to carry out the long division $$\begin{array}{r} x^2 - 5 \\ x+2 \overline{\smash{)}x^3 + 2x^2 - 5x + 7} \\ \underline{x^3 + 2x^2} \\ -5x - 10 \\ \underline{17} \end{array}$$ i.e. the quotient is $x^2 - 5$ and the remainder 17. We can check, $f(-2) = -8 + 8 + 10 + 7 = 17$.

Practice questions E

1 The expression $x^3 + ax^2 + bx - 3$ leaves a remainder of 7 when divided by $x - 2$ and a remainder of -20 when divided by $x + 1$. Calculate the remainder when the expression is divided by $x - 1$.

2 The expression $2x^3 - 3x^2 + bx + c$ leaves a remainder -6 when divided by $x - 1$ and has a factor $x + 2$.
Calculate the values of b and c.

3 The expression $x^3 + ax^2 + bx + c$ is divisible by both x and $x - 3$ but leaves a remainder of -40 when divided by $x + 2$. Determine

(a) the value of a, b, c

(b) all the factors of the expression

(c) the remainder when the expression is divided by $x - 1$.

4 (a) The expression $2x^3 + ax^2 + bx - 2$ leaves remainder of 7 and 0 when divided by $2x - 3$ and $x + 2$ respectively. Calculate the values of a and b.

(b) With these values of a and b factorise the expression completely.

5 The function $f(x) = 2x^3 + ax^2 + bx + 3$ has a factor $x + 3$. When $f(x)$ is divided by $x - 2$ the remainder is 15.

(a) Calculate the values of a and b.

(b) Find the other two factors of $f(x)$.

6 The expression $6x^2 + x + 7$ leaves the same remainder when divided by $x - a$ and by $x + 2a$, where $a \neq 0$. Calculate the value of a.

7 The polynomial g(x), where
$$g(x) \equiv 2x^3 + ax^2 - 3x + b,$$
has remainder 20 when divided by $(x - 2)$. The equation g(x) = 0 has a solution $x = 1$.

(a) Find the values of the constants a and b.

(b) Solve the equation g(x) = 0.

8 The remainder when $x^2 + 3x + 20$ is divided by $x - a$ is twice the remainder when it is divided by $x + a$. Find the possible values of a.

9 The remainder when $x^4 + 3x^2 - 2x + 2$ is divided by $x + a$ is the square of the remainder when $x^2 - 3$ is divided by $x + a$. Calculate the possible values of a.

10 The remainder when $a(a - b)(a + b)$ is divided by $a - 2b$ is $\frac{3}{4}$. Find the numerical value of b.
Using this value of b, find the remainder when $a(a - b)(a + b)$ is divided by $a + 2b$.

11 (a) When the expression $x^4 - ax^3 + a^2x^2 - 25a^4$ is divided by $x + 2a$, the remainder is 243. Find the possible values of a.

(b) The expression $x^2 + 7x + 3$ has the same remainder whether divided by $x - p$ or $x + q$, where $p \neq -q$. Find the value of $p - q$.

12 When a polynomial is divided by $(x - 2)$ the remainder is 2, and when it is divided by $(x - 3)$ the remainder is 5. When it is divided by $(x - 2)(x - 3)$ the remainder is $ax + b$. Find the values of a and b.

SUMMARY EXERCISE

There are no exam-style questions for the section on partial fractions: the technique is almost invariably coupled with an application to either a series, differentiation or integration and questions will appear for these in the relevant section.

In practice, questions on the remainder theorem normally involve parts on the factor theorem, factorisation etc. which was covered in P1, so some bits of these questions will be revision for you.

1 A quadratic polynomial $ax^2 + bx + c$ leaves remainder 1 when divided by $(x - 2)$ and remainder 2 when divided by $(x - 1)$.
If $ax^2 + bx + c \equiv a(x - 1)(x - 2) + px + q$,
find the numerical values of p and q.

2 The remainder when $2x^3 + ax^2 + bx - 3$ is divided by $x + 1$ is 4 and when divided by $x - 3$ is 84. Find the values of a and b and factorise the polynomial completely.

3 (a) Given that $f(x) = x^3 + kx^2 - 2x + 1$ and that when $f(x)$ is divided by $(x - k)$ the remainder is k, find the possible values of k.

(b) When the polynomial $p(x)$ is divided by $(x - 1)$ the remainder is 5 and when $p(x)$ is divided by $(x - 2)$ the remainder is 7. Given that $p(x)$ may be written in the form

$$(x - 1)(x - 2)q(x) + Ax + B$$

where $q(x)$ is a polynomial and A and B are numbers, find the remainder when $p(x)$ is divided by $(x - 1)(x - 2)$.

4 (a) Find, in terms of p, the remainder when

$$3x^3 - 2x^2 + px - 6 \text{ is divided by } x + 2.$$

Hence write down the value of p for which the expression is exactly divisible by $x + 2$.

(b) Solve the equation

$$x^3 - 12x + 16 = 0$$

(c) Given that the expression $x^3 + ax^2 + bx + c$ leaves the same remainder when divided by $x - 1$ or $x + 2$, prove that $a = b + 3$.

Given also that the remainder is 3 when the expression is divided by $x + 1$, calculate the value of c.

5 (a) The expression $x^3 + 2x^2 + ax + 4$ leaves a remainder of 10 when divided by $x + 3$. Determine the value of a and hence find the remainder when the expression is divided by $2x - 3$.

(b) Solve the equation

$$2x^3 + 5x^2 = 2 - x$$

(c) The expression $x^2 + ax + b$ leaves a remainder of p when it is divided by $x - 1$ and leaves a remainder of $p + 6$ when it is divided by $x - 2$. Find the value of a.

6 The polynomial $x^5 - 3x^4 - 2x^3 + 3x + 1$ is denoted by $f(x)$.

(a) Show that neither $(x - 1)$ nor $(x + 1)$ is a factor of $f(x)$

(b) By substituting $x = 1$ and $x = -1$ in the identity

$$f(x) \equiv (x^2 - 1) q(x) + ax + b,$$

where $q(x)$ is a polynomial and a and b are constants, or otherwise, find the remainder when $f(x)$ is divided by $(x^2 - 1)$

(c) Show, by carrying out the division, or otherwise, that when $f(x)$ is divided by $(x^2 + 1)$, the remainder is $2x$.

(d) Find all the real roots of the equation $f(x) = 2x$.

7 The polynomials $P(x)$ and $Q(x)$ are defined by

$$P(x) = x^8 - 1$$
$$Q(x) = x^4 + 4x^3 + ax^2 + bx + 5$$

(a) Show that $x - 1$ and $x + 1$ are factors of $P(x)$.

(b) It is known that when $Q(x)$ is divided by $x^2 - 1$ a remainder $2x + 3$ is obtained. Find the values of a and b.

(c) With these value of a and b find the remainder when the polynomial $[3P(x) + 4Q(x)]$ is divided by $x^2 - 1$.

SUMMARY

When you have finished, you should:

- know how to express rational functions in partial fractions when the denominator consists of

    (a) linear factors

    (b) one linear, one quadratic factor

    (c) one linear, one repeated (linear) factor

    (d) a polynomial of degree less than or equal to the degree of the numerator

- know the remainder theorem and applications.

P3 Section 1

ANSWERS

Practice questions A

1. (a) $\dfrac{1}{x+1} + \dfrac{2}{2x-1}$ (b) $\dfrac{3}{x+2} - \dfrac{1}{x+1}$

   (c) $\dfrac{2}{x+4} - \dfrac{1}{x+1}$ (d) $\dfrac{1}{x-3} - \dfrac{1}{x+1}$

   (e) $\dfrac{1}{x-2} - \dfrac{1}{x+2}$ (f) $\dfrac{2}{x-2} - \dfrac{3}{2x+1}$

   (g) $\dfrac{-2}{x} + \dfrac{1}{x+1} + \dfrac{1}{x-1}$

   (h) $\dfrac{1}{2(x-1)} - \dfrac{3}{2(x+1)}$

Practice questions B

1. (a) $\dfrac{-2}{x+2} + \dfrac{2x+1}{x^2+1}$ (b) $\dfrac{2}{2x+3} - \dfrac{3}{x^2+1}$

   (c) $\dfrac{-1}{x+1} + \dfrac{x+2}{x^2+1}$ (d) $\dfrac{1}{x} - \dfrac{x-3}{x^2+1}$

   (e) $\dfrac{1}{x} - \dfrac{x}{x^2+x+1}$ (f) $\dfrac{1}{x-2} - \dfrac{2x+1}{2x^2+3}$

   (g) $\dfrac{3}{x-2} + \dfrac{2-3x}{x^2+1}$

   (h) $\dfrac{1}{2(x-2)} - \dfrac{x}{2(x^2+4)}$

Practice questions C

1. (a) $\dfrac{2}{2-x} + \dfrac{2}{1+x} - \dfrac{1}{(1+x)^2}$

   (b) $\dfrac{1}{x} + \dfrac{1}{x-1} + \dfrac{3}{(x-1)^2}$

   (c) $\dfrac{1}{x+2} + \dfrac{1}{(x+2)^2} - \dfrac{2}{3x-1}$

   (d) $\dfrac{1}{x-1} - \dfrac{1}{(x-1)^2} - \dfrac{2}{2x+1}$

   (e) $\dfrac{-1}{x} - \dfrac{1}{x^2} + \dfrac{1}{x-1}$

   (f) $\dfrac{3}{x+1} + \dfrac{2}{(x+1)^2} - \dfrac{3}{x+2}$

   (g) $\dfrac{1}{x-1} + \dfrac{1}{(x-1)^2}$

   (h) $\dfrac{2}{x} + \dfrac{1}{x-1} + \dfrac{5}{(x-1)^2}$

Practice questions D

1. (a) $1 + \dfrac{4}{x+2} - \dfrac{9}{x+3}$

   (b) $1 + \dfrac{1}{x-1} - \dfrac{1}{x+1}$

   (c) $1 + \dfrac{3}{2(x-1)} - \dfrac{1}{2(x+1)}$

   (d) $x + 1 - \dfrac{1}{x} + \dfrac{2}{x-1}$

   (e) $x + \dfrac{1}{2(x-1)} + \dfrac{1}{2(x+1)}$

   (f) $1 + \dfrac{1}{5(x-3)} - \dfrac{1}{5(x+2)}$

   (g) $2x - 1 + \dfrac{1}{x+3} - \dfrac{3}{x-2}$

   (h) $2 + \dfrac{5}{x-3} + \dfrac{x}{x^2+1}$

2. $1, 2, -1$

Practice questions E

1. 4
2. $b = -11, c = 6$
3. (a) $-5, 6, 0$ (b) $x(x-2)(x-3)$ (c) 2
4. (a) $a = 3, b = -3$
   (b) $(x+2)(2x+1)(x-1)$
5. (a) $3, -8$ (b) $(2x-1)(x-1)$
6. $\dfrac{1}{6}$ 7. (a) $3, -2$ (b) $1, -\dfrac{1}{2}, -2$
8. $4, 5$ 9. $\dfrac{7}{9}, -1$ 10. $\dfrac{1}{2} : -\dfrac{3}{4}$
11. (a) ± 3 (b) -7
12. $3, -4$

SECTION 2
Coordinate geometry

INTRODUCTION We begin this section with an introduction to the geometry of a circle, its centre and radius. We then look at alternative ways of defining two related variables: sometimes it is more convenient to express the relationship between two variables with the aid of a third. This third variable is called the *parameter*. The two equations created (one linking the first variable to the parameter and the other linking the second variable to the parameter) are called *parametric* equations.

There are a number of standard curves with equations that can be conveniently expressed in parametric form: we shall have a look at these, and some additional curves, and see how to change from parametric form to the more familiar cartesian form and vice versa. Later on in this unit we shall be extending this, working out methods of differentiation and integration for curves in parametric form.

The equation of a circle
OCR P3 5.3.2 (c)

If a circle has centre (a, b) and radius r,

Figure 2.1

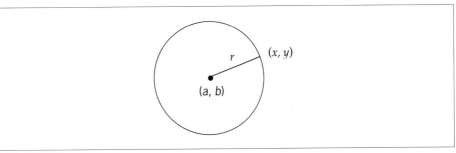

and if then (x, y) is any point on the circle, we have, using Pythagoras' theorem,

$$(x - a)^2 + (y - b)^2 = r^2$$

This is the general equation of a circle.

Example A circle has centre $(1, 3)$ and radius 5. Write down in expanded form its equation.

Solution We use the general equation with $a = 1$, $b = 3$ and $r = 5$. This gives

$$(x - 1)^2 + (y - 3)^2 = 5^2$$

Multiplying out the brackets we get:

$$x^2 - 2x + 1 + y^2 - 6y + 9 = 25$$
$$\therefore \quad x^2 + y^2 - 2x - 6y - 15 = 0.$$

13

Example	Write down the coordinates of the centre and the radius of the circle whose equation is $(x-3)^2 + (y+4)^2 = 36$.
Solution	Centre is $(3, -4)$. Radius $= \sqrt{36} = 6$.
Example	A circle has equation $x^2 + y^2 - 6x + 8y - 24 = 0$. Find its centre and radius.
Solution	Begin by rewriting the equation: $$x^2 - 6x + y^2 + 8y = 24$$ Now complete the square for each section: $$(x-3)^2 - 9 + (y+4)^2 - 16 = 24$$ Finally, rewrite: $$(x-3)^2 + (y+4)^2 = 49$$ ∴ Centre is $(3, -4)$ and radius $= \sqrt{49} = 7$.

Practice questions A

1 Find the centre and radius of the following circles.
  (a) $(x-2)^2 + (y+4)^2 = 25$
  (b) $(x+1)^2 + y^2 = 3$
  (c) $x^2 - 6x + y^2 - 10y - 2 = 0$
  (d) $x^2 + 2x + y^2 - 2y = 2$
  (e) $x^2 + y^2 = 10y$
  (f) $2x^2 + 2x + 2y^2 - 6y = 13$

2 Give the equation of a circle with
  (a) centre $(0, 0)$ radius 3
  (b) centre $(-1, -2)$, radius 5
  (c) centre $(-3, 4)$ radius 5
  (d) centre $(1, \sqrt{2})$ radius $\sqrt{3}$

  In each case, give also the expanded form of the equation (where possible).

Tangents to a circle

OCR P3 5.3.2 (d)

There are two ways of finding the gradient of a tangent to a circle. Either we use the fact that a tangent is at right-angles to the radius at the same point:

Figure 2.2

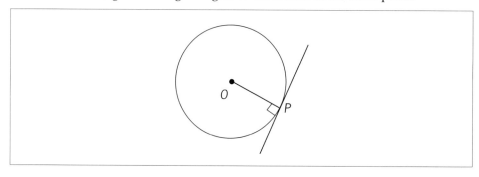

or we wait until we look at implicit differentiation in the calculus section later on in this unit! For the moment we can use the first method.

Example

Verify that the point $P(-2, -2)$ lies on the circle $x^2 + y^2 - 2x - 4y - 20 = 0$. What is the equation of the tangent at P?

Solution

We substitute the values $x = -2$ and $y = -2$ into the equation and show that the equation is satisfied

$$(-2)^2 + (-2)^2 - 2(-2) - 4(-2) - 20 = 0$$

∴ P lies on the circle.

To find the centre and radius, we complete the square for the terms in x and again for the terms in y. This gives

$$(x - 1)^2 + (y - 2)^2 = 5^2$$

i.e. circle centre $(1, 2)$, radius 5. We sketch the circle and find the gradient of the radius CP.

Figure 2.3

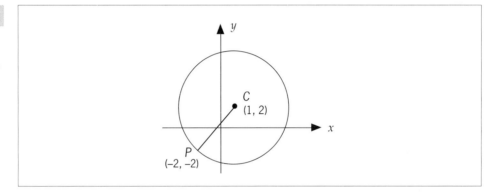

Gradient of $CP = \dfrac{2 - (-2)}{1 - (-2)} = \dfrac{4}{3}$

∴ gradient of tangent at P is $\dfrac{-3}{4}$

∴ equation of tangent at P is:

$$y + 2 = -\frac{3}{4}(x + 2) \quad \text{or} \quad 3x + 4y + 14 = 0$$

We can also find the length of the tangent from a point to a circle.

Example

Find the length of the tangent from the point $Q(5, 5)$ to the circle

$$x^2 + y^2 - 2x + 4y - 4 = 0$$

Solution

The equation of the circle can be rewritten as:

$$(x - 1)^2 + (y + 2)^2 = 9$$

∴ Centre is $C(1, -2)$ and radius is 3.

Figure 2.4

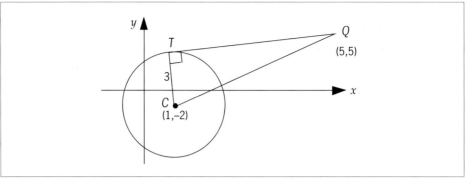

Now $CQ^2 = (5-1)^2 + (5+2)^2 = 65$

∴ Pythagoras on triangle CTQ gives
$TQ^2 = 65 - 3^2$

∴ length of tangent = $TQ = \sqrt{56}$.

Points of intersection

In general, a line will meet a circle at two distinct points, which we find by solving the equations simultaneously. For example, the line $y = x + 1$ meets the circle $x^2 + y^2 = 25$ at points whose x-values are given by
$$x^2 + (x+1)^2 = 25, \quad \text{i.e.} \quad x^2 + x - 12 = 0$$

This last quadratic has distinct roots $x = 3$ and $x = -4$ and so the two intersection points are $(3, 4)$ and $(-4, -3)$.

However, if a line is a tangent to a circle, then the resulting quadratic will have a repeated root, i.e. the discriminant will be zero: hence, the condition for repeated roots is
$$b^2 - 4ac = 0 \quad \text{or} \quad b^2 = 4ac.$$

A line is therefore a tangent to a circle if the resulting quadratic has repeated roots, i.e. $b^2 = 4ac$.

Example

The line $y = mx$ is a tangent to the circle $x^2 + y^2 - 2x + 6y + 5 = 0$.
Find the values of m and the possible contact points.

Solution

Substituting for y we get:

$$x^2 + (mx)^2 - 2x + 6(mx) + 5 = 0$$
∴ $(1 + m^2) x^2 + (6m - 2) x + 5 = 0$

Repeated roots ∴ $(6m - 2)^2 = 4(1 + m^2) 5$

∴ $2m^2 - 3m - 2 = 0$ ∴ $m = 2$ or $-\frac{1}{2}$

Now $m = 2 \Rightarrow 5x^2 + 10x + 5 = 0 \Rightarrow x = -1 \Rightarrow y = -2$

∴ contact point is $(-1, -2)$

and $m = -\frac{1}{2} \Rightarrow 1\frac{1}{4} x^2 - 5x + 5 = 0 \Rightarrow x = 2 \Rightarrow y = -1$

∴ other contact point is $(2, -1)$.

Practice questions B

1. Show that $P(7, -2)$ lies on the circle
$$x^2 + y^2 - 6x - 2y - 15 = 0$$
What is the equation of the tangent at P?

2. Find the length of the tangent from $Q(2, 1)$ to the circle:
$$x^2 + y^2 + 4x - 4y - 8 = 0$$

3. The line $y = mx$ is a tangent to the circle
$$x^2 + y^2 + 2x + 14y + 40 = 0.$$
Find the values of m and the corresponding contact points.

4. The line $y = mx + 3$ is a tangent to the circle
$$x^2 + y^2 + 2x - 12y + 32 = 0$$
Find the values of m and the corresponding contact points.

5. Show that the line $3y - x = 3$ is a tangent to the circle $x^2 + y^2 - 10x - 12y + 51 = 0$ and find the contact point.

6. If the line $y = mx + c$ is a tangent to the circle $x^2 + y^2 = a^2$, prove that $c^2 = a^2(1 + m^2)$.

7. A, B, C are the points $(-2, -4)$, $(3, 1)$, $(-2, 0)$. Find the equation of the circle passing through A, B, C and show that the tangent at B is parallel to the diameter through C.
(Hint: let the equation be
$$x^2 + y^2 + ax + by + c = 0$$
and find three simultaneous equations.)

Parametric equations

OCR P3 5.3.2 (a)

Suppose you ask two people to make up rules for transforming any number you care to give them.

One of them says: 'Double it and add 3.'

The other says: 'Multiply it by 3 and subtract 4.'

If t stands for your given number, and x and y for the results of the two transformations, then you can summarise what they said by writing:

$$x = 2t + 3 \quad \text{and} \quad y = 3t - 4$$

For example, if you choose '3', then $x = 2 \times 3 + 3 = 9$ and $y = 3 \times 3 - 4 = 5$

if you choose '1', then $x = 5$ and $y = -1$

In fact, by choosing different values of t, you can make x have any value you like, or y have any value you like – but not both at the same time. They are related together by the variable t and are not free to change independently of each other. When t is used like this to define the relation between two other variables, we call it a *parameter*.

The parametric equations give us an alternative way of describing the relationship between the variables x and y: sometimes one form is easier, say for sketching the corresponding curve, and sometimes the other, say for calculating associated areas using integration. We need to be able to change from one form to the other.

Finding the cartesian equation

OCR P3 5.3.2 (b)

If we want to find out what the direct relationship is between x and y, without the parameter acting as go-between, we need to eliminate the parameter.

This is usually possible, although we may need to use different methods depending on the particular pair.

In many cases you can do this by making the parameter t the subject of one of the parametric equations and substituting this into the other equation.

Example $x = 2t + 3$ and $y = 3t - 4$

Solution We can rearrange the first of these to give

$$t = \frac{x-3}{2}$$

and putting this into the second gives:

$$y = 3\left(\frac{x-3}{2}\right) - 4 = \frac{3x}{2} - \frac{9}{2} - 4 = \frac{3x}{2} - \frac{17}{2}$$

i.e. $2y = 3x - 17$

This is called the *cartesian equation*.

Example $x = t^2 - t$ and $y = 2t + 1$

Solution We can use the same method for this pair.

Here $y = 2t + 1$, so $t = \frac{y-1}{2}$

Putting this into the x-equation gives

$$x = \left(\frac{y-1}{2}\right)^2 - \left(\frac{y-1}{2}\right)$$

Example $x = 3 \cos \theta$ and $y = 3 \sin \theta$

Solution To find the cartesian equivalent of this pair, we need to use one of the pythagorean identities, $\cos^2 \theta + \sin^2 \theta = 1$.

Rearranging the equations, $\cos \theta = \frac{x}{3}$ and $\sin \theta = \frac{y}{3}$ and squaring, $\cos^2 \theta = \frac{x^2}{9}$ and $\sin^2 \theta = \frac{y^2}{9}$. Substituting into the identity, we have the cartesian equation

$$\frac{x^2}{9} + \frac{y^2}{9} = 1 \Rightarrow x^2 + y^2 = 9$$

i.e. a circle, centre (0, 0) and radius 3.

Example $x = 1 + \tan \theta$ and $y = 2 \sec \theta$

Solution We use the same method here, except that the identity is $1 + \tan^2 \theta = \sec^2 \theta$.

Rearranging, $\tan \theta = x - 1$, $\sec \theta = \frac{y}{2} \Rightarrow \tan^2 \theta = (x-1)^2$ and $\sec^2 \theta = \left(\frac{y}{2}\right)^2$

and the cartesian equation is

$$1 + (x-1)^2 = \left(\frac{y}{2}\right)^2$$

Example

$x = t + \dfrac{1}{t}$ and $y = t - \dfrac{1}{t}$

Solution

This uses a sort of trick. If you square both of the equations,

$$x^2 = \left(t + \frac{1}{t}\right)^2 = t^2 + 2 + \frac{1}{t^2}$$

$$y^2 = \left(t - \frac{1}{t}\right)^2 = t^2 - 2 + \frac{1}{t^2}$$

and subtract, $x^2 - y^2 = \left(t^2 + 2 + \dfrac{1}{t^2}\right) - \left(t^2 - 2 + \dfrac{1}{t^2}\right) = 4$

and so the cartesian equation is $x^2 - y^2 = 4$.

Example

$x = 2\cos\theta - \sin\theta$ and $y = 3\cos\theta + \sin\theta$

Solution

This uses another trick. The equations are first solved simultaneously to find $\cos\theta$ and $\sin\theta$ in terms of x and y and then fed into the identity.

$x = 2\cos\theta - \sin\theta$
$y = 3\cos\theta + \sin\theta$

Solving these simultaneously, $\cos\theta = \dfrac{1}{5}(x+y)$, $\sin\theta = \dfrac{1}{5}(2y - 3x)$.

Putting these into $\cos^2\theta + \sin^2\theta = 1$ gives the cartesian equation

$$\frac{1}{25}(x+y)^2 + \frac{1}{25}(2y - 3x)^2 = 1$$

which can be simplified by expanding, etc., to give $2x^2 - 2xy + y^2 = 5$

You would be given some hint in the exam for the last two types. The others, using the methods of direct substitution and trigonometric identities, you would be expected to be familiar with.

To reverse this process, i.e. to find suitable parametric equations for a given cartesian equation, we need either to recognise the type or be given some lead, generally in the form of one of the equations.

Example

Find suitable parametric equations for:

(a) $(x+4)^2 + (y-2)^2 = 25$ (d) $xy = c^2$

(b) $9x^2 + 4y^2 = 36$ (e) $y^2 = 4ax$

(c) $(y-1)^2 - (x+2)^2 = 1$

Solution

(a)   This is a circle. We are going to use $\cos^2\theta + \sin^2\theta = 1$, so we need to divide by 25 first of all:

$$\frac{(x+4)^2}{25} + \frac{(y-2)^2}{25} = 1$$

i.e. $\left(\dfrac{x+4}{5}\right)^2 + \left(\dfrac{y-2}{5}\right)^2 = 1$

So if we put $\frac{x+4}{5} = \cos\theta$ and $\frac{y-2}{5} = \sin\theta$, i.e.

$$x = 5\cos\theta - 4 \text{ and } y = 5\sin\theta + 2$$

we have suitable equations.

(b) We use the same identity for this equation, which represents an ellipse. Dividing by 36,

$$\frac{9x^2}{36} + \frac{4y^2}{36} = 1 \Rightarrow \left(\frac{x}{2}\right)^2 + \left(\frac{y}{3}\right)^2 = 1$$

$$\Rightarrow \frac{x}{2} = \cos\theta, \frac{y}{3} = \sin\theta \text{ and our equations are}$$

$$x = 2\cos\theta, y = 3\sin\theta$$

(c) Here the identity is $\sec^2\theta - \tan^2\theta = 1$

and so $\quad y - 1 = \sec\theta, x + 2 = \tan\theta.$

i.e. $\quad x = \tan\theta - 2, y = \sec\theta + 1$

(d) This is called a *rectangular hyperbola*: there are many possible pairs for this, but the generally accepted one is

$$x = ct, y = \frac{c}{t}$$

(e) This is a parabola, with standard parametric equations

$$x = at^2, y = 2at$$

Example Given the cartesian equation $y = x\sqrt{4-x^2}$, find a suitable parametric equation for y corresponding to the parametric equation $x = 2\cos\theta$.

Solution Substituting the parametric equation into the cartesian,

$$y = 2\cos\theta\sqrt{4 - 4\cos^2\theta} = 2\cos\theta\sqrt{4(1-\cos^2\theta)}$$

$$= 2\cos\theta\sqrt{4\sin^2\theta} = 2 \times 2\cos\theta\sin\theta$$

$$= 2\sin 2\theta$$

The parametric equations are $x = 2\cos\theta, y = 2\sin 2\theta$

Practice questions C

1 Find the cartesian equation for the following pairs of parametric equations.

(a) $x = t + 1, y = t^2 + t$ (b) $x = t^2, y = t^3$

(c) $x = \frac{8}{t}, y = 4t$ (d) $x = 2at, y = \frac{a}{t}$

(e) $x = t - 1, y = \frac{t+2}{t+3}$

(f) $x = at^2, y = at^3$ (a constant)

(g) $x = 2\cos\theta, y = 2\sin\theta$

(h) $x = 1 + \cos\theta, y = 3 + \sin\theta$

(i) $x = 3 - 2\cos\theta, y = 5 + 2\sin\theta$

(j) $x = 3\cos\theta, y = 2\sin\theta$

(k) $x = \tan\theta - 2, y = 1 + \sec\theta$

(l) $x = \cos 2\theta, y = \sin\theta$

2 Find suitable parametric equations for
(a) $x^2 + y^2 = 49$ (b) $\frac{x^2}{16} + \frac{y^2}{25} = 1$
(c) $(x+2)^2 + (y-1)^2 = 9$ (d) $y^2 - 4x^2 = 1$
(e) $xy = 16$ (f) $y^2 = x^3$
(g) $y^2 = 8x$ (h) $y = 2x^2 - 1$

3 If the parametric equations for x and y are
$$x = 4\left(t + \frac{1}{t}\right), \ y = 3\left(t - \frac{1}{t}\right)$$
find and simplify expressions for $3x + 4y$ and $3x - 4y$.
Hence find the cartesian equation of the curve.

4 Find the coordinates of the points where the curve with parametric equations
$$x = t^2 + 2t, \ y = 3t - 1$$
meets the y-axis.

5 The curve with parametric equations
$$x = 2 \cos \theta, \ y = 3 - \sin \theta,$$
cuts the y-axis at A and B. Find the length AB.

6 If $x = \operatorname{cosec} \theta - \cot \theta$ and $y = \operatorname{cosec} \theta - 2 \cot \theta$, express $\operatorname{cosec} \theta$ and $\cot \theta$ in terms of x and y. Hence give an expression for y in terms of x only.

7 [Harder] Find the cartesian equations for the following pairs of parametric equations
(a) $x = \sec \theta - \tan \theta, \ y = \sec \theta + 2 \tan \theta$
(b) $x = \frac{t}{t+1}, \ y = \frac{t}{t+2}$
(c) $x = \frac{t}{1+t}, \ y = \frac{t^2}{1+t}$
(d) $x = (t-1)^2, \ y = t^2 - 1$
(e) $x = \tan \theta, \ y = \cos 2\theta$

8 If $x = t^3 - t$ and $y = t^2 + t$, express $\frac{x}{y}$ as a function of t in its simplest form and hence find the cartesian equation.

Curve sketching

OCR P3 5.3.2 (a)

Standard curves

There are just a few of these that you really need to know

(a) $x = at, \ y = \frac{a}{t}$

This is a rectangular hyperbola: the corresponding cartesian equation is $xy = a^2$ or $y = \frac{a^2}{x}$. The curve looks like this:

Figure 2.5

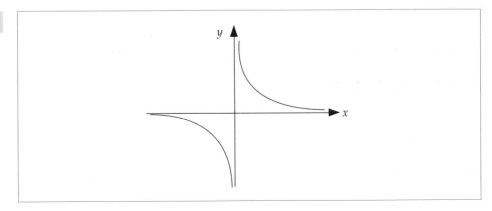

(b) $x = at^2$, $y = 2at$

This is a parabola: eliminating t gives the equation $y^2 = 4ax$ with curve:

Figure 2.6

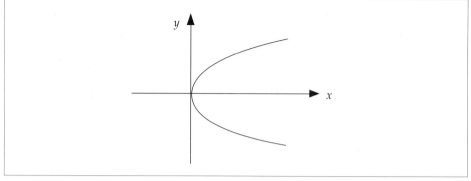

(c) $x = at^2$, $y = at^3$

This is more complicated: x cannot be negative but y can be, y increases more rapidly than x. The cartesian equation is $ay^2 = x^3$ and the curve looks something like this:

Figure 2.7

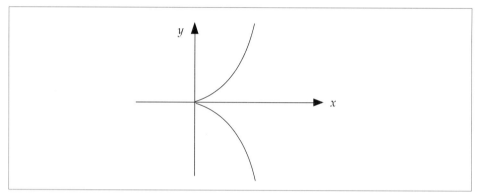

The shape at the origin where the two parts converge with a gradient of zero is called a **cusp**.

(d) $x = a \cos t$, $y = a \sin t$ $0 \leq t < 2\pi$

This is the standard parametric equation of a circle, centre origin and radius a, i.e. with cartesian equation $x^2 + y^2 = a^2$

Figure 2.8

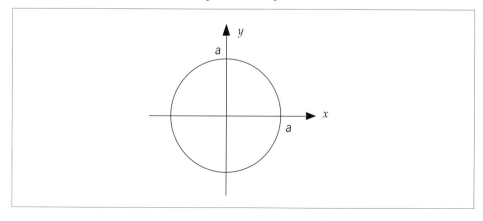

(e) $x = a \cos t,\ y = b \sin t,\ 0 \leq t < 2\pi$

If $a \neq b$, this is an ellipse with standard cartesian equation $\dfrac{x^2}{a^2} + \dfrac{y^2}{b^2} = 1$

Figure 2.9

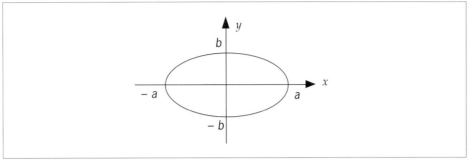

The final curve in this series you are not expected to know, but it has a nice shape and demonstrates how parametric equations can make life easier.

(f) $x = a \cos^3 t,\ y = a \sin^3 t$

Figure 2.10

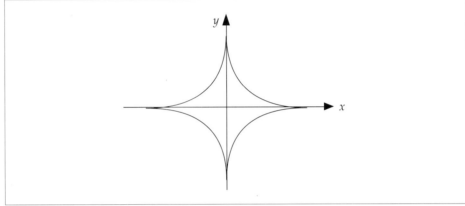

The cartesian equation is $x^{\frac{2}{3}} + y^{\frac{2}{3}} = a^{\frac{2}{3}}$

Besides these standard curves, you may be asked to sketch a curve with equation in parametric form: there are two possible ways of doing this.

One of these is to look at the parametric equations, see what information they give, take some values for the parameter to get some particular points and plot these.

Example

Sketch the curve given by the parametric equations

$x = t^2 - 1$ and $y = 2t + 1$

Solution

We'll find first of all the points where the curve crosses the axes. This will be when $x = 0 \Rightarrow t^2 - 1 = 0 \Rightarrow t = 1$ and $t = -1$.

Putting these values in for y gives $y = 3$ and $y = -1$, giving the points $(0, 3)$ and $(0, -1)$

$y = 0 \Rightarrow 2t + 1 = 0 \Rightarrow t = -\dfrac{1}{2}$.

At this point, $x = (-\dfrac{1}{2})^2 - 1 = -\dfrac{3}{4}$, giving the point $(-\dfrac{3}{4}, 0)$

Looking at the function which defines x, we can see that x is an even function – there are two values of t (each giving a different value for y) which give the same value for x (for example, x is 8 for both t = 3 and t = −3). Also, the value of x can't be less than −1, and it reaches this minimum value when t = 0, in which case y = 1. It also increases in value much quicker than y, which is plodding along as a linear function to x's quadratic function. If we find a couple more points we can make a reasonable attempt at sketching the curve: t = 3 gives (8,7) and t = −3 gives (8,−5), so joining all these points we have:

Figure 2.11

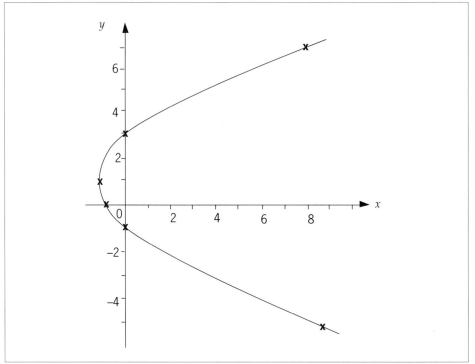

The other method is to convert the parametric equations into cartesian form and work from this. So for example above, finding t from the y-equation, $t = \frac{y-1}{2}$ and into the x-equation, $x = \left(\frac{y-1}{2}\right)^2 - 1$. (This is actually a complicated transformation of $x = y^2$: translation + 1 y-direction, stretch factor 2 x-direction, translation −1 x-direction.)

Practice questions D

1 Sketch the following graphs
 (a) $x = 3 \cos t, y = 3 \sin t; \ 0 \leq t < 2\pi$
 (b) $x = 2t, y = \frac{2}{t}$
 (c) $x = t^2, y = t^3$
 (d) $x = 4 \cos t, y = 2 \sin t; \ 0 \leq t < 2\pi$
 (e) $x = 2t^2, y = 4t$
 (f) $x = 1 + 3 \cos t, y = 2 + 3 \sin t; 0 \leq t < 2\pi$

2 Sketch the following, using values in the parametric equations
 (a) $x = t^2 - 1, y = t - 1$
 (b) $x = 2t - 3, y = t + 1$
 (c) $x = \frac{1}{t+1}, y = \frac{t}{t+1}$
 (d) $x = \frac{1}{t+1}, y = t + 2$

Transformations of the curve $y = \dfrac{1}{x}$

OCR P4 5.4.1 (b)

To conclude this section, we look at some members of the family of curves called rectangular hyperbolas, with shape as you probably know.

Figure 2.12

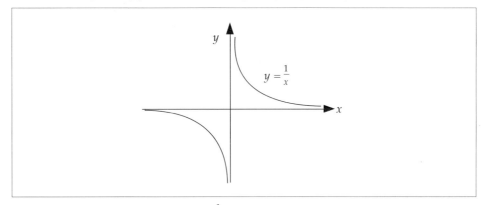

Any equation of the form $y = \dfrac{ax + b}{cx + d}$ is going to be a variant on this basic configuration: we can either work out the successive transformations or find a few points and deduce the shape from these.

Example

Sketch the curve

$$y = \dfrac{x}{x - 1}$$

Solution

If we divide, $y = 1 + \dfrac{1}{x-1}$ and we can see the transformations quite easily: translation +1 in x direction, translation +1 in y-direction. These translations shift the asymptotes that were the x- and y-axes, and the resulting curve looks like this:

Figure 2.13

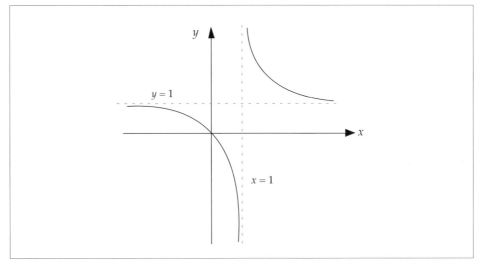

Alternatively, we look at the bottom of the fraction, $x - 1$. This cannot be zero, so we have the x-asymptote at $x = 1$.

As x gets larger positively, y tends towards 1 from above $\left(\text{e.g. } \dfrac{1000}{1000 - 1}\right)$, so the y-asymptote is $y = 1$.

Finally, when $x = 0$, $y = 0$.

With these particular pieces of information and knowing the general shape, we can assemble the curve above.

Example

Sketch the curve $y = \dfrac{x}{2 + x}$

Solution

Dividing, $y = 1 - \dfrac{2}{2 + x}$

Translation -2 x-direction

Stretch 2 y-direction, reflection in x-axis

Translation $+1$ y-direction

Figure 2.14

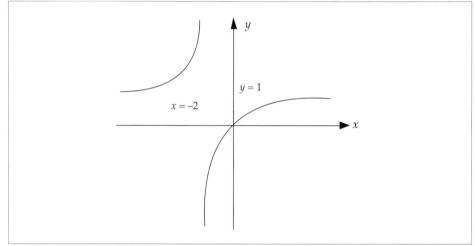

Alternatively;   bottom of fraction, x-asymptote $x = -2$

x large $+ve \Rightarrow y \to 1$ from below $\left(\dfrac{1000}{1002}\right)$

i.e. y-asymptote at $y = 1$

$x = 0, y = 0$

and this is enough to sketch the curve.

Practice questions E

1   Sketch the curves

    (a) $y = \dfrac{x+1}{x-1}$ (b) $y = \dfrac{x}{x+1}$ (c) $y = \dfrac{x-2}{x+2}$ (d) $y = \dfrac{3-x}{x+2}$

SUMMARY EXERCISE

1 A circle with centre C has equation
$$x^2 + y^2 - 6x + 4y = 7.$$
  (a) Determine the coordinates of C and the radius of the circle
  (b) Find the x-coordinates of the points A and B where the circle cuts the x-axis.
  (c) Hence obtain the cosine of angle ABC, leaving your answer in surd form.
  (This uses the fact that
  $$\cos A\hat{B}C = \frac{AB^2 + BC^2 - AC^2}{2 \cdot AB \cdot BC},$$
  the cosine-rule.) [AQA 1999]

2 The circle with equation $x^2 + y^2 - 4x + 6y = 12$ has centre C.
  (a) Determine the coordinates of C and calculate the radius of the circle.
  (b) The points P and Q lie on the circumference of the circle such that the arc length PQ is equal to 10 units. Calculate the area of the sector PCQ. [AQA 1998]

3 A circle has equation $x^2 + y^2 + 2x - 8y = 152$
  (a) Find the radius and the coordinates of the centre.
  (b) The point P lies on the circle and has coordinates $(k, 16)$ where $k > 0$.
    (i) Determine the value of k.
    (ii) Find the coordinates of the point Q at the opposite end of the diameter from P. [AQA 1995]

4 Determine the coordinates of the centre C and the radius of the circle with equation
$$x^2 + y^2 + 4x - 6y = 12$$
The circle cuts the x-axis at the points A and B. Calculate the area of the triangle ABC.
Calculate the area of the minor segment of the circle cut off by the chord AB, giving your answer to three significant figures. [AQA]

5 The equation of a circle is $x^2 + y^2 - 8y = 9$. Find the coordinates of the centre of the circle, and the radius of the circle.

A straight line has equation $x = 3y + k$, where k is a constant. Show that the y-coordinates of the points of intersection of the line and the circle are given by
$$10y^2 + (6k - 8)y + (k^2 - 9) = 0$$
Hence determine the exact values of k for which the line is a tangent to the circle.

6 The points P and Q have coordinates $(3, 2)$ and $(1, 6)$ respectively.
  (a) Determine a cartesian equation for the straight line through P and Q.
  (b) The circle with equation
  $$x^2 + y^2 - 8x + ay + b = 0$$
  passes through the points P and Q.
    (i) Calculate the value of a and the value of b.
    (ii) State the coordinates of the centre of this circle and determine its radius. [AQA 1998]

7 The points P, Q on the rectangular hyperbola $xy = c^2$ have coordinates $(cp, \frac{c}{p})$ and $(cq, \frac{c}{q})$ respectively. Find and simplify the gradient of the chord PQ and deduce that the tangent at P has equation
$$p^2 y + x = 2cp$$

8 The points O, A, B, C have coordinates $(0, 0)$, $(1, 0)$, $(1, 1)$, $(0, 1)$, respectively, and P is the point (x, y). Write down, in terms of x and y, an expression for d^2, where
$$d^2 = PO^2 + PA^2 + PB^2 + PC^2,$$
and show that
$$d^2 = 4[(x - \tfrac{1}{2})^2 + (y - \tfrac{1}{2})^2 + \tfrac{1}{2}]$$
  (a) Deduce that no point P exists for which $d^2 < 2$
  (b) Given that $d^2 = 2$, find the coordinates of P
  (c) Given that $d^2 = 6$, show that the possible positions of P lie on a circle and give the centre and radius of this circle.

9 With respect to a fixed origin O, the points A and B have coordinates $(-4, 0)$ and $(12, 12)$ respectively. The mid-point of AB is M. Find an equation of the line in the plane of the coordinate axes Ox and Oy which passes through M and is perpendicular to AB.

Hence, or otherwise, find, in cartesian form, an equation of the circle which passes through O, A and B.

10 $f(x) \equiv \dfrac{p - 2x}{x + q}$, $x \in \mathbb{R}$, $X \neq -q$,

where p and q are constants. The curve, C, with equation $y = f(x)$ has an asymptote with equation $x = 2$ and passes through the point with coordinates (3, 2).

(a) Write down the value of q

(b) Show that $p = 8$

(c) Write down an equation of the second asymptote to C

(d) Using $p = 8$ and the value of q found in (a), sketch the graph of C showing clearly how C approaches the asymptotes and the coordinates of the points where C intersects the axes.

11 Sketch the curve whose equation is

$$y = \dfrac{2x + 3}{x + 1}.$$

The curve cuts the x-axis at A and the y-axis at B, and the asymptotes to the curve intersect at C. Find the coordinates of C.

12 Sketch the curve whose equation is

$$y = \dfrac{x - 2}{x + 2}$$

and state the equations of its asymptotes.

On the same diagram sketch the curve whose equation is

$$\dfrac{x^2}{4} + y^2 = 1$$

Hence, or otherwise, find all real solutions of the equation

$$4(x - 2)^2 = (4 - x^2)(x + 2)^2$$

SUMMARY

You should now:

- be familiar with the general equation for a circle

- be able to find the centre and radius of a circle given the equation in expanded form

- be able to find the equations of tangents to circles

- be able to find intersections of lines and circles

- know that parametric equations is an alternative form of presenting a relation between two variables

- be able to change from cartesian to parametric forms and vice versa

- be familiar with the shape of some standard curves in parametric equations

- know how to sketch a curve with equation in parametric form

- be able to sketch curves with equation of the type $y = \dfrac{ax + b}{cx + d}$.

ANSWERS

Practice questions A

1. (a) (2, –4); 5   (b) (–1, 0); √3
   (c) (3, 5); 6   (d) (–1, 1); 2
   (e) (0, 5): 5   (f) $\left(\frac{-1}{2}, \frac{3}{2}\right)$: 3

2. (a) $x^2 + y^2 = 9$ (b) $(x + 1)^2 + (y + 2)^2 = 25$
   $\qquad x^2 + 2x + y^2 + 4y = 20$
   (c) $(x + 3)^2 + (y - 4)^2 = 25$
   $\qquad x^2 + 6x + y^2 - 8y = 0$
   (d) $(x - 1)^2 + (y - \sqrt{2})^2 = 3$
   $\qquad x^2 - 2x + y^2 - 2\sqrt{2}y = 0$

Practice questions B

1. $3y = 4x - 34$
2. $\sqrt{33}$
3. $m = \frac{13}{9}$ or -3 : contact points $\left(\frac{-18}{5}, \frac{-26}{5}\right)$, (2, –6)
4. $m = 2$ or $-\frac{1}{2}$: (1, 5) and (–2, 4)
5. (6, 3)
7. $x^2 + y^2 - 2x + 4y - 8 = 0$

Practice questions C

1. (a) $y = (x - 1)^2 + (x - 1)$
   (b) $y^2 = x^3$ (c) $xy = 32$
   (d) $xy = 2a^2$ (e) $y = \frac{x + 3}{x + 4}$
   (f) $ay^2 = x^3$ (g) $x^2 + y^2 = 4$
   (h) $(x - 1)^2 + (y - 3)^2 = 1$
   (i) $(x - 3)^2 + (y - 5)^2 = 4$
   (j) $\frac{x^2}{9} + \frac{y^2}{4} = 1$ (k) $1 + (x + 2)^2 = (y - 1)^2$
   (l) $x = 1 - 2y^2$

2. (a) $x = 7 \cos \theta, y = 7 \sin \theta$
   (b) $x = 4 \cos \theta, y = 5 \sin \theta$
   (c) $x = 3 \cos \theta - 2, y = 3 \sin \theta + 1$
   (d) $x = \frac{1}{2} \tan \theta, y = \sec \theta$
   (e) $x = 4t, y = \frac{4}{t}$
   (f) $x = t^2, y = t^3$
   (g) $x = 2t^2, y = 4t$
   (h) Many, e.g. $x = t, y = 2t^2 - 1$ or $x = \cos \theta$, $y = \cos 2\theta$

3. $24t, \frac{24}{t}$: $9x^2 - 16y^2 = 576$
4. (0, –1), (0, –7)
5. 2
6. $\operatorname{cosec} \theta = 2x - y$, $\cot \theta = x - y$
   $\Rightarrow 1 + (x - y)^2 = (2x - y)^2$
7. (a) $x(x + 2y) = 3$ (b) $2y = x + xy$
   (c) $y = x^2 + xy$ (d) $(y - x)^2 = 4x$
   (e) $y = \frac{1 - x^2}{1 + x^2}$
8. $\frac{x}{y} = t - 1$: $y^3 = (x + y)(x + 2y)$

Practice questions D

1. (a) 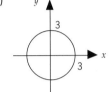   (b)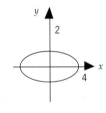

   (c) 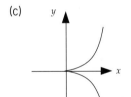   (d)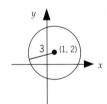

   (e) 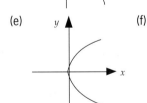   (f)

2. (a) 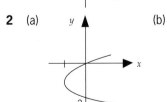   (b)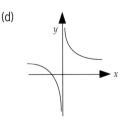

   (c) 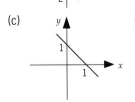   (d)

Practice questions E

1. (a) 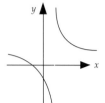 (b)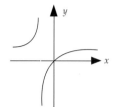

   (c) 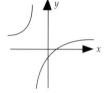 (d)

SECTION 3
Series

INTRODUCTION In the previous unt we looked at the expansion of $(a + b)^n$, where n was a positive integer, which gave a finite number of terms, $n + 1$ in fact. We have a different situation when n is rational – the resulting series is infinite.

This section looks at the general binomial series and how this can be applied to find expressions which can be used as approximations to rational functions by taking only the first few terms.

General binomial expansion

OCR P3 5.3.1 (c)

The reason that we want to turn rational functions into series is similar to the reasoning behind turning fractions (which are exact and relatively short) into decimals (which can go on for ever and are usually only approximations). It is that decimals have a standard form – their value can be seen immediately and compared with other known values and they can also be added, multiplied, etc. more easily. Although they are not precise, for most purposes two or three places of decimals are sufficient.

It's a very similar process with rational functions and roots – provided certain conditions are met,

$$\frac{1}{1-x} = 1 + x + x^2 + x^3 + x^4 + \dots .$$

In fact, if we substitute $x = \frac{1}{10}$ we have

$$\frac{1}{1-\frac{1}{10}} = \frac{10}{9} = 1 + \frac{1}{10} + \frac{1}{100} + \frac{1}{1000} + \dots$$

$$= 1 \cdot 1\,1\,1\,1 \cdots \dots$$

Just as three places of decimals are usually enough, so that $\frac{10}{9} \approx 1.111$, only terms with powers of x lower than or equal to 3 need usually be included, i.e.

$$\frac{1}{1-x} \approx 1 + x + x^2 + x^3$$

It's the same with square roots, $\sqrt{3}$ is actually 1.7320508 … but for most purposes we take it to be 1.732.

$\sqrt{1+x}$, if it's expressed as an infinite series, would start

$$1 + \frac{1}{2}x - \frac{1}{8}x^2 + \frac{1}{16}x^3 + \dots$$

and we could take this to be $1 + \frac{1}{2}x - \frac{1}{8}x^2$ the majority of the time.

When you have understood this the next task is to discover where these infinite series of terms come from and how to work out their value.

In fact, there is very little difference between working out the terms for an infinite series and the finite series that we've already looked at. We can't use the ^nC_r function because n is going to be either a fraction or a negative number, and anyway we won't be needing any but the first few terms.

Let's write out some of the expansion of $(1 + x)^n$, without using the ^nC_r function:

$$(1 + x)^n = 1 + \frac{n}{1} \cdot x + \frac{n(n-1)}{2.1} x^2 + \frac{n(n-1)(n-2)}{3.2.1} x^3 + \ldots$$

This is true for any value of n – integer, fraction, positive or negative (but there sometimes has to be a restriction on the value of x: this restriction is that $-1 < x < 1$, i.e. $|x| < 1$. This ensures that the terms become smaller in the same way that for a GP, the common ratio r has to satisfy the condition $|r| < 1$.) Some examples of this are:

$$n = \tfrac{1}{2}: \quad (1+x)^{\frac{1}{2}} = 1 + \left(\tfrac{1}{2}\right)x + \frac{\left(\tfrac{1}{2}\right)\left(-\tfrac{1}{2}\right)}{2.1} x^2 + \frac{\left(\tfrac{1}{2}\right)\left(-\tfrac{1}{2}\right)\left(-\tfrac{3}{2}\right)}{3.2.1} x^3$$

$$= 1 + \frac{x}{2} - \frac{x^2}{8} + \frac{x^3}{16} + \ldots$$

$$n = -1: \quad (1+x)^{-1} = 1 + (-1)x + \frac{(-1)(-2)}{2.1} x^2 + \frac{(-1)(-2)(-3)}{3.2.1} x^3 + \ldots$$

$$= 1 - x + x^2 - x^3 + \ldots$$

To avoid making errors try and get into the habit of writing expansions down in this fixed pattern. As before, if the x is replaced by $2x$ or $-x$, the expansion is basically unchanged.

$$(1+2x)^{\frac{1}{2}} = 1 + \left(\tfrac{1}{2}\right)(2x) + \frac{\left(\tfrac{1}{2}\right) \cdot \left(-\tfrac{1}{2}\right)}{2.1}(2x)^2 + \frac{\left(\tfrac{1}{2}\right) \cdot \left(-\tfrac{1}{2}\right) \cdot \left(-\tfrac{3}{2}\right)}{3.2.1}(2x)^3 + \ldots$$

$$= 1 + x - \frac{x^2}{2} + \frac{x^3}{2} - \ldots$$

and

$$(1-x)^{-2} = 1 + (-2)(-x) + \frac{(-2)(-3)(-x)^2}{2.1} + \frac{(-2)(-3)(-4)(-x)^3}{3.2.1} + \ldots$$

$$= 1 + 2x + 3x^2 + 4x^3 + \ldots$$

Practice questions A

1 Write out the first four terms in the expansion of :

(a) $(1+x)^{\frac{1}{3}}$ (b) $(1-x)^{\frac{1}{2}}$ (c) $(1+2x)^{-1}$

(d) $(1-3x)^{\frac{2}{3}}$ (e) $(1-2x)^{-\frac{1}{2}}$ (f) $\left(1+\tfrac{x}{2}\right)^{-2}$

simplifying the coefficients as far as possible.

Note that the examples above are written in the most convenient form for expansion – you could well find them disguised as:

(a) $\sqrt[3]{1+x}$ (b) $\sqrt{1-x}$ (c) $\frac{1}{1+2x}$

(d) $\sqrt[3]{(1-3x)^2}$ (e) $\frac{1}{\sqrt{1-2x}}$ (f) $\frac{1}{\left(1+\tfrac{x}{2}\right)^2}$

2 Show that $\dfrac{1}{\sqrt{1-x}} - \sqrt{1+x} = \dfrac{x^2}{2} + \dfrac{x^3}{4}$

if x^4 and higher powers of x may be neglected.

3 If x is small, show that:

$$(1-x^2)^{\frac{-1}{3}} \approx 1 + \tfrac{1}{3}x^2 + \tfrac{2}{9}x^4$$

Multiple series

OCR P3 5.3.1 (c)

We do this quite simply multiplying term by term and then collecting, for example,

$$\begin{aligned}(1 - x + x^2)(1 + 3x + 4x^2) &= 1 + 3x + 4x^2 \\ & - x - 3x^2 - 4x^3 \\ & + x^2 + 3x^3 + 4x^4 \\ \hline &= 1 + 2x + 2x^2 - x^3 + 4x^4\end{aligned}$$

Usually, however, the series are short because terms with powers of x higher than 2 can be ignored. This means an easier job multiplying out as we only keep terms up to and including x^2. Let's have a look at an example of this.

Example Express $\frac{1+x}{1-x}$ as an expression up to and including the term in x^2.

Solution Rewriting this as $(1+x)(1-x)^{-1}$, we need to find the first three terms of $(1-x)^{-1}$

$$(1-x)^{-1} = 1 + (-1)(-x) + \frac{(-1)(-2)(-x)^2}{2 \cdot 1}$$
$$= 1 + x + x^2$$

So $\frac{1+x}{1-x} = (1+x)(1+x+x^2) = 1 + x + x^2 + x + x^2$
$$= 1 + 2x + 2x^2$$

when we ignore any terms with powers of x greater than 2.

We'll try another of these in the next example.

Example Express $\sqrt{\left(\frac{1-2x}{1+2x}\right)}$ as a series in ascending powers of x, up to and including the term in x^2.

Solution We can rewrite the fraction as

$$(1-2x)^{\frac{1}{2}}(1+2x)^{-\frac{1}{2}}$$

and expand each of these separately.

$$(1-2x)^{\frac{1}{2}} = 1 + \left(\frac{1}{2}\right)(-2x) + \frac{\left(\frac{1}{2}\right)\left(-\frac{1}{2}\right)(-2x)^2}{2}$$
$$= 1 - x - \frac{x^2}{2}$$

$$(1+2x)^{-\frac{1}{2}} = 1 + \left(-\frac{1}{2}\right)(2x) + \frac{\left(-\frac{1}{2}\right)\left(-\frac{3}{2}\right)(2x)^2}{2}$$
$$= 1 - x + \frac{3x^2}{2}$$

Then

$$\sqrt{\left(\frac{1-2x}{1+2x}\right)} = \left(1 - x - \frac{x^2}{2}\right)\left(1 - x + \frac{3x^2}{2}\right) = 1 - x + \frac{3x^2}{2} - x + x^2 - \frac{x^2}{2}$$
$$= 1 - 2x + 2x^2, \text{ up to and including the term in } x^2.$$

Practice questions B

1. Express as a series in ascending powers of x, up to and including the term in x^2:

   (a) $\dfrac{1+2x}{1-2x}$ (b) $\dfrac{1-x}{\sqrt{1+x}}$

   (c) $\dfrac{(1+x)^2}{(1-x)^2}$ (d) $\sqrt[3]{\left(\dfrac{1-3x}{1+3x}\right)}$

2. Determine the value of the constant k given that in the expansion of
$$\frac{1-kx^2}{(1-x^2)^{\frac{1}{2}}}$$
in ascending powers of x for $|x|<1$, the coefficient of x^2 is zero. Using this value of k, find the first three non-zero terms of the expansion.

3. Show that if x is so small that terms in x^3 and higher powers may be neglected, then
$$\left(\frac{1-x}{1+x}\right)^{\frac{1}{4}} = 1 - \frac{1}{2}x + \frac{1}{8}x^2$$

4. Show that the first three terms in the expansion in ascending powers of x of
$$(1+8x)^{\frac{1}{4}}$$
are the same as the first three terms in the expansion of
$$\frac{1+5x}{1+3x}$$

Rearrangements

OCR P3 5.3.1 (c)

In a way, the general expansion that has been quoted is quite limited – it only works when the terms inside the bracket start with 1. To expand something like
$$\sqrt{4+x}$$
we have to rearrange it somewhat to bring the bracket into the standard form

$$\sqrt{4+x} = (4+x)^{\frac{1}{2}} = \left[4\left(1+\frac{x}{4}\right)\right]^{\frac{1}{2}}$$

$$= 4^{\frac{1}{2}}\left(1+\frac{x}{4}\right)^{\frac{1}{2}} = 2\left(1+\frac{x}{4}\right)^{\frac{1}{2}}$$

$$= 2\left[1 + \left(\tfrac{1}{2}\right)\left(\tfrac{x}{4}\right) + \frac{\left(\tfrac{1}{2}\right)\left(-\tfrac{1}{2}\right)\left(\tfrac{x}{4}\right)^2}{2}\right]$$

$$= 2\left[1 + \frac{x}{8} - \frac{x^2}{128}\right]$$

$$= 2 + \frac{x}{4} - \frac{x^2}{64}$$

up to and including the term in x^2.

(If a bracket has been rearranged into the form $a(1+bx)^n$, so that the leading term is 1, the condition that the series is valid is that $|bx|<1$. You don't need to know this, but it explains why in questions involving the binomial expansion, you will see $|x|<\frac{1}{2}$ or something similar.)

This procedure of rearranging, or lack of it, is a very common source of error. Be particularly careful when bringing brackets up from the bottom.

$$\frac{1}{2-x} = (2-x)^{-1}$$

$$= \left[2\left(1-\frac{x}{2}\right)\right]^{-1}$$

$$= 2^{-1}\left(1-\frac{x}{2}\right)^{-1}$$

$$= \frac{1}{2}\left[1 + (-1)\left(-\frac{x}{2}\right) + \frac{(-1)(-2)\left(-\frac{x}{2}\right)^2}{2}\right]$$

$$= \frac{1}{2}\left[1 + \frac{x}{2} + \frac{x^2}{4}\right]$$

$$= \frac{1}{2} + \frac{x}{4} + \frac{x^2}{8} \text{ neglecting powers of } x \geq 3.$$

Practice questions C

1 Put the following into the form $a(1 + bx)^n$. The expansion is not required

(a) $(8 + x)^{\frac{1}{3}}$ (b) $\dfrac{1}{x+2}$

(c) $\dfrac{1}{\sqrt{9-x}}$ (d) $\dfrac{1}{(27+9x)^{\frac{2}{3}}}$

2 Find a quadratic approximation to

$\sqrt{\left(\dfrac{4-x}{1-2x}\right)}$, given that $|x| < \frac{1}{2}$

3 If x is so small that terms in x^n, $n \geq 3$ can be neglected and

$$\frac{3 + ax}{3 + bx} = (1-x)^{\frac{1}{3}}$$

find the values of a and of b.

Larger terms

OCR P3 5.3.1 (c)

You occasionally come across questions where instead of the bracket containing something like $1 + 2x$, it contains something like $1 + x + x^2$. There is no change in the form of the expansion. $(2x)^3$ would be replaced by $(x + x^2)^3$ and the terms become a little more complicated to work out. Let's have a look at one of these.

Example

Expand $(1 + x + x^2)^{\frac{1}{2}}$ as a series in ascending powers of x up to and including the term in x^3.

Solution

Expanding as normal,

$$(1 + x + x^2)^{\frac{1}{2}} = 1 + \frac{1}{2}(x + x^2) + \frac{\left(\frac{1}{2}\right)\left(-\frac{1}{2}\right)(x + x^2)^2}{2!}$$

$$+ \frac{\left(\frac{1}{2}\right)\left(-\frac{1}{2}\right)\left(-\frac{3}{2}\right)(x + x^2)^3}{3!}$$

$$= 1 + \frac{1}{2}x + \frac{1}{2}x^2 - \frac{1}{8}(x + x^2)^2 + \frac{1}{16}(x + x^2)^3$$

(We can expand these brackets, but don't need to take all the terms.)

$$= 1 + \frac{1}{2}x + \frac{1}{2}x^2 - \frac{1}{8}\left[x^2 + 2x^3 + \ldots\right] + \frac{1}{16}\left[x^3 + \ldots\right]$$

$$= 1 + \frac{1}{2}x + \frac{1}{2}x^2 - \frac{1}{8}x^2 - \frac{x^3}{4} + \frac{x^3}{16}$$

$$= 1 + \frac{1}{2}x + \frac{3}{8}x^2 - \frac{3}{16}x^3 \text{ up to and including } x^3$$

Similarly, if the bracket contains something like $1 + \frac{2}{x}$, the series will be in ascending powers of $\frac{1}{x}$. For example

$$\left(1 + \frac{2}{x}\right)^{\frac{1}{2}} = 1 + \frac{1}{x} - \frac{1}{2x^2} + \frac{1}{2x^3} - \ldots$$

Practice questions D

1 Expand $(1 + 2x - x^2)^{-1}$ as a series in ascending powers of x, up to and including the term in x^3.

2 Expand $\left(1 - \frac{2}{x}\right)^{\frac{1}{2}}$ as a series in ascending powers of $\frac{1}{x}$, up to and including the term in $\frac{1}{x^2}$.

3 If x^4 and higher powers of x can be neglected, show that

$$\sqrt{\left(\frac{1-x}{1+x+x^2}\right)} = 1 - x + \frac{1}{2}x^3$$

Use of partial fractions

OCR P3 5.3.1 (c)

When we have a more complicated algebraic fraction that we would like to express as an ascending series in x, our work can be made easier by the use of partial fractions. Suppose, for example, that we wanted to find the first four non-zero terms in the expansion in ascending powers of x of:

$$\frac{1 - 8x}{(1 + x)(1 - 2x)}$$

In partial fractions $\dfrac{1 - 8x}{(1 + x)(1 - 2x)} \equiv \dfrac{A}{1 + x} + \dfrac{B}{1 - 2x}$

$$= \frac{A(1 - 2x) + B(1 + x)}{(1 + x)(1 - 2x)}$$

$$1 - 8x \equiv A(1 - 2x) + B(1 + x)$$

$x = \frac{1}{2}$ $\quad -3 = 0 + B\left(\frac{3}{2}\right) \quad \Rightarrow B = -2$

$x = -1$ $\quad 9 = A(3) + 0 \quad \Rightarrow A = 3$

i.e. $\dfrac{1 - 8x}{(1 + x)(1 - 2x)} \equiv \dfrac{3}{1 + x} - \dfrac{2}{1 - 2x}$

So to find the series for the left-hand side, we find each of the series for the right-hand side and subtract.

$$\frac{3}{1+x} = 3(1+x)^{-1} = 3\left[1 - x + x^2 - x^3 + \ldots\right]$$

$$\frac{2}{1-2x} = 2(1-2x)^{-1} = 2\left[1 + 2x + 4x^2 + 8x^3 + \ldots\right]$$

so finally

$$\frac{1-8x}{(1+x)(1-2x)} = 3\left[1 - x + x^2 - x^3\right] - 2\left[1 + 2x + 4x^2 + 8x^3\right]$$

$$= 1 - 7x - 5x^2 - 19x^3 \text{ (first four terms)}$$

Practice questions E

1. Assuming that x is sufficiently small that terms in x^3 and higher power can be neglected, show, by putting $f(x)$ into partial fractions and then expanding the result, that

   (a) $\dfrac{1}{(1+x)(2+x)} \approx \dfrac{1}{8}(4 - 6x + 7x^2)$

   (b) $\dfrac{(1-x)^2}{(1+x)(1+x^2)} = 1 - 3x + 3x^2$

   (c) $\dfrac{x^2 + 5x}{(1+x)(1-x)^2} \approx 5x + 6x^2$

   (d) $\dfrac{2 + x + 2x^2}{(2+x)(1+x^2)} \approx 1$

   (e) $\dfrac{2 + 5x + 15x^2}{(2-x)(1+2x^2)} \approx 1 + 3x + 7x^2$

2. Express $f(x) = \dfrac{2 + 11x}{(2+x)(1-2x)}$ in partial fractions.

   Hence, or otherwise, determine the coefficient of x^3 in the expansion of $f(x)$ in a series of ascending powers of x.

3. Express the function

   $$\frac{1 + 3x^2}{(1-x)^2(1+x)}$$

   as the sum of three partial fractions. Hence, or otherwise, find the first three terms in the expansion of the function in ascending powers of x.

4. Show that if x is so small that x^4 and higher powers can be neglected then

   $$\frac{1 + 2x + 3x^2}{(1-x)(1+x^2)}$$

   can be expressed in the form $A + Bx + Cx^2 + Dx^3$ and find A, B, C and D.

Use in approximation

OCR P3 5.3.1 (c)

As we saw previously with the finite expansions, we can use a small number of terms with a suitable value of x to provide an approximation.

Example

Find the first four terms in the expansion of $(1-x)^{\frac{1}{2}}$ and use this with the value of $x = 0.02$ to find the value of $\sqrt{2}$ to 4 significant figures.

Solution

$$(1-x)^{\frac{1}{2}} = 1 + \frac{1}{2}(-x) + \frac{\frac{1}{2}\left(-\frac{1}{2}\right)(-x)^2}{2!}$$

$$= 1 - \frac{1}{2}x - \frac{1}{8}x^2$$

When $x = 0.02$, this series has the value

$$1 - \frac{1}{2}(0.02) - \frac{1}{8}(0.02)^2$$

$$= 0.98995 \qquad \ldots \text{①}$$

Substituting $x = 0.02$ into the bracket gives

$$(1 - 0.02)^{\frac{1}{2}} = \sqrt{0.98} = \sqrt{\left(\frac{98}{100}\right)} = \sqrt{\left(\frac{49}{100} \times 2\right)} = \frac{7}{10}\sqrt{2} \quad \ldots \text{②}$$

Equating ① and ②.

$$0.98995 = \frac{7}{10}\sqrt{2} \Rightarrow \sqrt{2} = 1.414 \text{ (to 4 s.f.)}$$

Practice questions F

1. Find the first four terms of the expansion of $\sqrt{1 + 2x}$. By putting $x = \frac{1}{100}$, find $\sqrt{102}$ correct to 4 decimal places.

2. Show that if x is so small that terms in x^3 and higher powers can be neglected,
$$\sqrt{\left(\frac{1+x}{1-x}\right)} = 1 + x + \frac{1}{2}x^2$$
By substituting $x = \frac{1}{9}$, show that $\sqrt{5} \approx \frac{181}{81}$.

SUMMARY EXERCISE

1. Given that $|x| < \frac{1}{2}$, expand $\sqrt{(1 + 2x)}$ as a series of ascending powers of x, up to and including the term in x^3, simplifying the coefficients.

2. Find, in ascending powers of x up to and including the term in x^3, the series expansion of $(4 + x)^{-\frac{1}{2}}$, giving your coefficients in their simplest form.

3. Use the binomial expansion to find $(1.0006)^{\frac{1}{3}}$ correct to eight places of decimals.

4. Find the first three terms in the series expansion, in ascending powers of x, of $(1 + x)^{-2}$, where $|x| < 1$.
Hence, or otherwise, show that, when x is small,
$$\left(\frac{1-x}{1+x}\right)^2 \approx 1 - 4x + 8x^2.$$

5. The binomial expansion of $(1 + 12x)^{\frac{3}{4}}$, in ascending powers of x as far as the term in x^3, is
$$1 + 9x + px^2 + qx^3 + \ldots, \ |12x| < 1.$$
   (a) Find the values of the constants p and q.

   (b) Use this expansion and your values of p and q to obtain an approximation to $(1.12)^{\frac{3}{4}}$. Give your answer to four decimal places.

6. Expand
   (a) $\sqrt[3]{(1 + px)}$, $|px| < 1$
   (b) $\dfrac{1 + 2qx}{1 + qx}$, $|qx| < 1$

   where p and q are constants, in terms of ascending powers of x up to and including the terms in x^2.

   Given that these terms of the expansions are the same, show that $p = 3q$.

7. $f(x) \equiv \dfrac{5 + x}{(1 + 2x)(1 - x)^2}$.
   (a) Express $f(x)$ in partial fractions.
   (b) Given that $|x| < \frac{1}{2}$, expand $f(x)$ in ascending powers of x, up to and including the term in x^3.

8 Obtain the expansion in ascending powers of x, up to and including the term in x^3, of
$$\frac{1+5x}{(1+2x)^{\frac{1}{2}}} \quad (|x|<\tfrac{1}{2}).$$
By putting $x = 0.04$ deduce an approximate value of $\frac{1}{\sqrt{3}}$, giving your answer to three decimal places. [AQA(AEB) 1994]

9 The binomial expansion of $(8 + x)^{\frac{1}{3}}$ in ascending powers of x, as far as the terms in x^2, is
$$(8 + x)^{\frac{1}{3}} = 2 + px + qx^2 + \ldots, \ |x| < 8$$
(a) Determine the values of the constants p and q.
(b) Use the expression $2 + pq + qx^2$, and your values of p and q, to obtain an estimate for $\sqrt[3]{15}$, giving your answer to 3 significant figures.
(c) Find the percentage error involved in using this estimate.

10 Find the values of the constants a and b for which the expansions, in ascending powers of x, of the two expressions $(1 + 2x)^{\frac{1}{2}}$ and $\frac{1+ax}{1+bx}$, up to and including the term in x^2, are the same.

With these values of a and b, use the result $(1+2x)^{\frac{1}{2}} \approx \frac{1+ax}{1+bx}$, with $x = -\frac{1}{100}$, to obtain an approximate value for $\sqrt{2}$ in the form p/q, where p and q are positive integers.

11 (a) Express
$$\frac{1-2x+5x^2}{(1-2x)(1+x^2)}$$
in the form
$$\frac{A}{1-2x} + \frac{Bx+C}{1+x^2}$$
for some constants A, B and C.

(b) Write down the series expansions, up to and including the terms in x^3, of $(1-2x)^{-1}$ and $(1+x^2)^{-1}$. Deduce that, if x is small enough for terms in x^4 and higher powers of x to be neglected, then
$$\frac{1-2x+5x^2}{(1-2x)(1+x^2)} = 1 + ax^2 + bx^3$$
for some constants a and b and state their values. [AQA(AEB) 1994]

12 $f(x) \equiv \dfrac{9-3x-12x^2}{(1-x)(1+2x)}$

Given that $f(x) \equiv A + \dfrac{B}{1-x} + \dfrac{C}{1+2x}$,

(a) find the values of the constants A, B and C.

(b) Given that $|x| < \frac{1}{2}$, expand $f(x)$ in ascending powers of x up to and including the term x^3, simplifying each coefficient.

SUMMARY

When you have finished this section, you should:

- be aware of the use of series as approximations to rational functions
- be familiar with the general expansion of $(1 + x)^n$ and be able to use it
- know how to multiply two series, neglecting terms above a certain power
- be able to use the revised series with a certain value for x to find approximations
- know how to rearrange $(a + bx)^n$ into the form $d(1 + cx)^n$

ANSWERS

Practice questions A

1. (a) $1 + \dfrac{x}{3} - \dfrac{x^2}{9} + \dfrac{5x^3}{81}$

   (b) $1 - \dfrac{x}{2} - \dfrac{x^2}{8} - \dfrac{x^3}{16}$

   (c) $1 - 2x + 4x^2 - 8x^3$

   (d) $1 - 2x - x^2 - \dfrac{4}{3}x^3$

   (e) $1 + x + \dfrac{3x^2}{2} + \dfrac{5x^3}{2}$

   (f) $1 - x + \dfrac{3x^2}{4} - \dfrac{x^3}{2}$

Practice questions B

1. (a) $1 + 4x + 8x^2$ (b) $1 - \dfrac{3x}{2} + \dfrac{7x^2}{8}$

   (c) $1 + 4x + 8x^2$ (d) $1 - 2x + 2x^2$

2. $\dfrac{1}{2} : 1 + \dfrac{x^4}{8} + \dfrac{x^6}{8}$

Practice questions C

1. (a) $2\left(1 + \dfrac{x}{8}\right)^{\frac{1}{3}}$ (b) $\dfrac{1}{2}\left(1 + \dfrac{x}{2}\right)^{-1}$

   (c) $\dfrac{1}{3}\left(1 - \dfrac{x}{9}\right)^{-\frac{1}{2}}$ (d) $\dfrac{1}{9}\left(1 + \dfrac{x}{3}\right)^{\frac{-2}{3}}$

2. $2 + \dfrac{7x}{4} + \dfrac{175}{64}x^2$

3. $a = -2$, $b = -1$

Practice questions D

1. $1 - 2x + 5x^2 - 12x^3$

2. $1 - \dfrac{1}{x} - \dfrac{1}{2x^2}$

Practice questions E

2. $\dfrac{-4}{2+x} + \dfrac{3}{1-2x} : \dfrac{97}{4}$

3. $\dfrac{-2}{1-x} + \dfrac{2}{(1-x)^2} + \dfrac{1}{1+x} : 1 + x + 5x^2$

4. $A = 1, B = 3, C = 5$ and $D = 3$

Practice questions F

1. 10.0995

SECTION 4

Differentiation

INTRODUCTION This is one of the biggest sections in this module and will extend greatly the expressions that we can differentiate; these include the trigonometric functions, products and quotients of functions, together with more complicated functions. We shall then look at methods for finding the gradient function when y is not simply given as a function of x: when it is given implicitly or when both the variables are given in terms of a parameter. You will then have quite a comprehensive basis for this important topic

Trigonometric functions
OCR P3 5.3.4 (a)

These have the following derivatives:

$y = \sin x$ then $\dfrac{dy}{dx} = \cos x$

$y = \cos x$ then $\dfrac{dy}{dx} = -\sin x$

$y = \tan x$ then $\dfrac{dy}{dx} = \sec^2 x$

The angle x must be in radians.

As they stand we cannot make much use of them, so we are going to look at combinations of functions, products and quotients and how these are differentiated.

Differentiation of a product
OCR P3 5.3.4 (b)

To differentiate a product, we use the following formula:

> If u and v are functions of x, and $y = uv$,
>
> then $\dfrac{dy}{dx} = u\dfrac{dv}{dx} + v\dfrac{du}{dx}$

So that if $y = x^2 \sin x$,

$\dfrac{dy}{dx} = x^2 \cos x + 2x \sin x$

and if $y = \ln x \tan x$, then

$\dfrac{dy}{dx} = \ln x \sec^2 x + \dfrac{1}{x} \tan x$

Note that since the two terms are added, it doesn't matter which part you differentiate and which you keep constant first of all.

Practice questions A

1 Differentiate the following with respect to x
(a) $x^3 \ln x$ (b) $x^2 e^x$ (c) $\sin x \cos x$ (d) $e^x \cos x$ (e) $x \tan x$
(f) $\sin x \ln x$ (g) $x^4 \sin x$ (h) $3 \cos x \ln 5x$

Differentiation of a quotient

OCR P3 5.3.4 (b)

There is a corresponding formula for quotients:

If u and v are functions of x and $y = \dfrac{u}{v}$

then $\dfrac{dy}{dx} = \dfrac{v \dfrac{du}{dx} - u \dfrac{dv}{dx}}{v^2}$

So if $y = \dfrac{\sin x}{x^2}$, $\dfrac{dy}{dx} = \dfrac{x^2 \cos x - \sin x \cdot 2x}{x^4}$

$= \dfrac{x \cos x - 2 \sin x}{x^3}$

Note that the negative sign between the terms on top of the fraction means that order is important here. It also means that you have to be a little careful with signs sometimes, giving each term its own bracket. Let's look at an example of this.

Example Find $\dfrac{dy}{dx}$ when $y = \dfrac{1 - x^2}{x^2}$

Solution $\dfrac{dy}{dx} = \dfrac{x^2(-2x) - (1 - x^2)2x}{x^4}$

$= \dfrac{-2x^3 - 2x + 2x^3}{x^4} = \dfrac{-2x}{x^4} = \dfrac{-2}{x^3}$

It's very easy if you don't put the terms in brackets to miss one of the minus signs.

Practice questions B

1 Differentiate the following with respect to x
(a) $\dfrac{x^2}{1-x}$ (b) $\dfrac{1 + \sin x}{1 + \cos x}$ (c) $\dfrac{\ln x}{x^2}$
(d) $\dfrac{\tan x}{e^x}$ (e) $\dfrac{x}{x^2 + 1}$ (f) $\dfrac{1 + e^x}{1 - e^x}$
(g) $\dfrac{e^x}{x^2}$ (h) $\dfrac{\sin x}{\cos x}$

2 Given that $y = \dfrac{\ln x}{x^2}$, find the value of $\dfrac{dy}{dx}$ at the point where $x = e$.

3 Find the turning points on the curve with equation
$$y = \dfrac{x^2}{1 + x^4}$$

4 The tangent to the curve $y = \dfrac{3 \tan x}{1 + \sin x}$ at the point where $x = \dfrac{\pi}{6}$ cuts the x-axis at T. Show that the distance from T to the origin is $\dfrac{1}{6}(2\sqrt{3} - \pi)$

5 Given that $y = \dfrac{x^2 - 1}{2x^2 + 1}$, find $\dfrac{dy}{dx}$ and state the values of x for which y is an increasing function.

Implicit differentiation

OCR P3 5.3.4 (d)

The functions we've differentiated so far have given one of the variables, usually y, explicitly in terms of the other variable, usually x, in the form of $y = f(x)$. When y is mixed in with the x's, given implicitly, we have to expand our technique of differentiation so that we can still find the gradient function at any point.

When we talk about differentiating some function, x^3, we assume that the differentiation is with respect to the same variable, x, and write:

$$\frac{d(x^3)}{dx} = 3x^2$$

If we want to differentiate the same function with respect to another variable, t say, we differentiate as before with respect to x, but then multiply by x differentiated with respect to t:

$$\frac{d(x^3)}{dt} = \frac{d(x^3)}{dx} \times \frac{dx}{dt} = 3x^2 \frac{dx}{dt}$$

You can see that it looks as though the dx's 'cancel' in the middle expression to give the left-hand side. Let's take another example and differentiate $\ln u$ with respect to v:

$$\frac{d(\ln u)}{dv} = \frac{d(\ln u)}{du} \times \frac{du}{dv} = \frac{1}{u} \frac{du}{dv}$$

We'll use this to find $\dfrac{dy}{dx}$ from an expression where the x's and y's are mixed together, such as:

$$x^3 + 3y^2 - 2x - 5y = 3$$

If we differentiate term by term with respect to x, which we should now be able to do without too much trouble, we get the following:

$$3x^2 + 6y \frac{dy}{dx} - 2 - 5 \frac{dy}{dx} = 0$$

Rearranging gives: $6y \dfrac{dy}{dx} - 5 \dfrac{dy}{dx} = 2 - 3x^2$

Factorising gives: $(6y - 5) \dfrac{dy}{dx} = 2 - 3x^2$

Dividing gives: $\dfrac{dy}{dx} = \dfrac{2 - 3x^2}{6y - 5}$

This is quite straightforward provided we remember:

(a) to put $\dfrac{dy}{dx}$ after differentiating a y term, and

(b) that constants differentiated are zero.

But when there are mixtures of x's and y's in the same term, for example $3x^3y^2$ it is necessary to differentiate using the product rule

$$\dfrac{d(3x^3y^2)}{dx} = \underset{\text{constant}}{3x^3} \times \underset{\text{differentiated}}{y^2} + \underset{\text{differentiated}}{3x^3} \times \underset{\text{constant}}{y^2}$$

$$= 3x^3 \times 2y\dfrac{dy}{dx} + 9x^2y^2$$

$$= 6x^3y\dfrac{dy}{dx} + 9x^2y^2$$

Differentiation of a^x

To find the derivative of $y = a^x$, where a is constant, we take lns to give

$$\ln y = \ln a^x = x \ln a$$

then we differentiate with respect to x; since $\ln a$ is constant, $x \ln a$ just differentiates to $\ln a$

$$\dfrac{1}{y}\dfrac{dy}{dx} = \ln a$$

Multiplying by y,

$$\dfrac{dy}{dx} = y \ln a = a^x . \ln a \quad \text{since } y = a^x.$$

This is another standard result which you are expected to remember.

$$\text{If } y = a^x \Rightarrow \dfrac{dy}{dx} = a^x . \ln a$$

So if $y = 3^x$, $\dfrac{dy}{dx} = 3^x . \ln 3$ for example.

Note that in the particular case where $a = e$,

$$y = e^x \Rightarrow \dfrac{dy}{dx} = e^x . \ln e = e^x \text{ (since } \ln e = 1)$$

which is the result we already knew.

Practice questions C

1 Differentiate the following expressions with respect to x
(a) y^3 (b) $\sin y$ (c) $\ln y$
(d) e^y (e) $3 \tan y$ (f) $-\cos y$

2 Using the product rule, differentiate the following expressions with respect to x
(a) x^2y^2 (b) xe^y (c) $x^3 \cos y$
(d) $y \sin x$ (e) $4xy^3$ (f) $\tan x \tan y$
(g) $(x+y)y^2$ (h) e^{x+y}

3 Find an expression for $\frac{dy}{dx}$ for the following

(a) $x^3 + y^3 = 5$ (b) $x^2 - y^2 = x$
(c) $y^2 + y = \sin x$ (d) $3x^2y^2 - 4xy^3 = 7$
(e) $(x+y)y = x^4$ (f) $\sin x + \sin y = \tan x$

4 If $x^3 + y^3 + 3xy - 1 = 0$, find the gradient of the curve at the point $(2, -1)$.

5 Find the gradient of the curve with equation $x^3 + y^3 = 9$ at the point $(1, 2)$.

6 If $\tan x + \tan y = 3$, find the value of $\frac{dy}{dx}$ when $x = \frac{\pi}{4}$.

7 If $\ln y = y \ln x$, find $\frac{dy}{dx}$ in terms of x and y.

8 Find the gradient function for the curve given by
$$x^3 + 2y^3 = 3x^2y^2$$
Hence find the coordinates of the point other than the origin where $\frac{dy}{dx} = 0$.

9 Differentiate with respect to x,

(a) 5^x (b) $\left(\frac{1}{2}\right)^x$

A function of a function

OCR P2 5.2.6 (b)

When we looked at implicit differentiation we found that we could differentiate a function of one variable with respect to another variable, for example:

$$\frac{d(u^3)}{dx} = 3u^2 \frac{du}{dx}$$

This is going to be useful now, because we can use this to differentiate something like:

$$y = (1 + x^2)^3$$

which at the moment we can't do directly. If we change the variable, however, we can temporarily simplify this expression. Instead of the $1 + x^2$, we put u so that now $y = u^3$. □This is called *making the substitution* $u = 1 + x^2$.

Differentiating both sides with respect to x gives:

$$\frac{dy}{dx} = 3u^2 \frac{du}{dx}$$

as we saw above. But if $u = 1 + x^2$, $\frac{du}{dx} = 2x$, so we can write this:

$$\frac{dy}{dx} = 3u^2 \times 2x$$
$$= 3(1 + x^2)^2 \times 2x$$
$$= 6x(1 + x^2)^2$$

which is once again in terms of x as we wanted. This is called the *chain-rule*: not a very long chain in this case, (only two variables) but there can be more.

Here are some further examples involving short chains.

Example Differentiate:

(a) $\ln(1 + 2x^3)$ (b) $\sin(x^2 - 1)$

Solution

(a) Put $y = \ln(1 + 2x^3)$ and $u = 1 + 2x^3$ so that:

$y = \ln u$

Differentiating with respect to x,

$$\frac{dy}{dx} = \frac{1}{u} \times \frac{du}{dx}$$

But $u = 1 + 2x^3$, so $\frac{du}{dx} = 6x^2$

and $\frac{dy}{dx} = \frac{1}{u} \times 6x^2 = \frac{6x^2}{1 + 2x^3}$

(b) Putting $y = \sin(x^2 - 1)$ and $u = x^2 - 1$, we have:

$y = \sin u$, so that

$$\frac{dy}{dx} = \cos u \times \frac{du}{dx}$$

$\frac{du}{dx} = 2x$ and $\frac{dy}{dx} = \cos u \times 2x$

$= 2x \cos(x^2 - 1)$

Most of the differentiation we meet with will involve expressions of this type and we need to be able to deal with them quickly and accurately. In practice, we only use this change of variable with complicated expressions involving long chains; for example, to differentiate $\ln\left(1 + \sqrt{\tan 4x}\right)$. For standard function of a function expressions there is a simpler way: here are two examples from which we can derive a general method.

(a) If $y = e^{5x}$; put $u = 5x$ so that $y = e^u$

then $\frac{dy}{dx} = e^u \times \frac{du}{dx}$: $u = 5x \Rightarrow \frac{du}{dx} = 5$

$= e^u \times 5$

$= 5e^{5x}$

(b) If $y = e^{\sin x}$; put $u = \sin x$ so that $y = e^u$

then $\frac{dy}{dx} = e^u \times \frac{du}{dx}$: $u = \sin x \Rightarrow \frac{du}{dx} = \cos x$

$= e^u \times \cos x$

$= \cos x \, e^{\sin x}$

Using $\rightarrow$ to mean differentiation with respect to x, we could write these two as:

$e^{5x} \rightarrow 5e^{5x}$

$e^{\sin x} \rightarrow \cos x e^{\sin x}$

In general we could write this as

$e^{\text{function}} \rightarrow \text{function differentiated} \times e^{\text{function}}$

If we write the function as □ and the derivative of the function as ■, we represent the rule as

$$e^{\square} \to \blacksquare e^{\square}$$

Using this as a model, we can write some of the simpler examples of this type straight down, for example:

$$e^{3x^2} \to 6xe^{3x^2}$$
$$e^{1-x} \to -e^{1-x}$$
$$e^{\tan x} \to \sec^2 x \, e^{\tan x}$$

Using the same shorthand, we can draw up another table for the standard functions.

$$e^{\square} \to \blacksquare e^{\square}$$
$$\sin \square \to \blacksquare \cos \square$$
$$\cos \square \to -\blacksquare \sin \square$$
$$\tan \square \to \blacksquare \sec^2 \square$$
$$\ln \square \to \frac{\blacksquare}{\square}$$

so that, for example

$$e^{5x} \to 5e^{5x}$$
$$\sin(x^2) \to 2x \cos(x^2)$$
$$\cos(e^x) \to -e^x \sin(e^x)$$
$$\tan 4x \to 4 \sec^2 4x$$
$$\ln(1 + 3x^2) \to \frac{6x}{1 + 3x^2}$$

Practice questions D

1 Write down the derivative with respect to x of the following

(a) $\sin(e^x)$ (b) e^{-4x} (c) $\ln(1 + 4x)$ (d) $\tan(x^3)$ (e) $\cos 3x$

(f) $e^{\cos x}$ (g) $\ln(1 + \sin x)$ (h) $\sin(\ln x)$ (i) $\cos\left(\frac{\pi}{2} - x\right)$ (j) $\tan\left(3x + \frac{\pi}{4}\right)$

One relationship that you have to know is that

$$\frac{dx}{dy} = \frac{1}{\frac{dy}{dx}}$$

Example

Find $\frac{dy}{dx}$ when $y = e^x + 1$.

Rearrange the equation to find x in terms of y, find $\frac{dx}{dy}$ and verify that $\frac{dx}{dy} = \frac{1}{\frac{dy}{dx}}$

Solution

$\frac{dy}{dx} = e^x$

If $y = e^x + 1 \Rightarrow e^x = y - 1$

$\Rightarrow x = \ln(y - 1)$

$\frac{dx}{dy} = \frac{1}{y-1} = \frac{1}{(e^x + 1) - 1} = \frac{1}{e^x}$

$= \frac{1}{\frac{dy}{dx}}$ as required.

Powers of a function

OCR P2 5.2.6 (b)

We can use the chain-rule to differentiate an expression involving a power of a function. For example, to differentiate $y = (1 + \sin x)^5$ we could substitute $u = 1 + \sin x$, and then $y = u^5$.

Since $\frac{du}{dx} = \cos x$, we have

$\frac{dy}{dx} = \frac{d(u^5)}{dx} = \frac{d(u^5)}{du} \times \frac{du}{dx} = 5u^4 \times \cos x$

$= 5(1 + \sin x)^4 \cos x$

this is true in general: if $y = [f(x)]^n$, then $\frac{dy}{dx} = n[f(x)]^{n-1} \times \frac{e\,f(x)}{dx}$, or using the symbols we used before, we can add a further expression to the table

$$\square^n \rightarrow n\square^{n-1} \blacksquare$$

This is a very useful expression and you will use it time and time again. (A very common mistake is to forget to multiply by ■, the derivative of the function in the brackets.)

Example

Differentiate the following with respect to x

(a) $(1 + x^2)^{10}$

(b) $\sqrt{1 + e^{4x}}$

(c) $\sin^3 2x$

(d) $\frac{1}{(1 - 2x)^3}$

Solution

(a) The rule in words is: bring down the power, reduce the power by one and multiply by the derivative of the bracket. This gives
$$(1 + x^2)^{10} \to 10(1 + x^2)^9 (2x) = 20x(1 + x^2)^9$$

(b) $\sqrt{1 + e^{4x}} = (1 + e^{4x})^{\frac{1}{2}} \to \frac{1}{2}(1 + e^{4x})^{-\frac{1}{2}}(4e^{4x})$

$$= 2e^{4x}(1 + e^{4x})^{-\frac{1}{2}}$$

$$= \frac{2e^{4x}}{\sqrt{1 + e^{4x}}}$$

(c) Remember that $\sin^3 2x = (\sin 2x)^3$

$\sin^3 2x = (\sin 2x)^3 \to 3(\sin 2x)^2 (2\cos 2x)$

$$= 6\sin^2 2x \cos 2x$$

(d) $\dfrac{1}{(1 - 2x)^3} = (1 - 2x)^{-3} \to -3(1 - 2x)^{-4}(-2)$

$$= 6(1 - 2x)^{-4}$$

$$= \frac{6}{(1 - 2x)^4}$$

Practice questions E

1 Write down and simplify as far as possible the derivative with respect to x of

(a) $(1 + 4x)^5$ (b) $(3 - e^x)^6$ (c) $\dfrac{1}{1 + 2x}$ (d) $\dfrac{1}{\sqrt{1 - x^2}}$ (e) $\cos^4 x$

(f) $\sqrt{\sin x}$ (g) $(1 + \tan x)^5$ (h) $(\sin x + \cos x)^4$ (i) $\tan^2 3x$ (j) $\sqrt{1 - 4x^3}$

(k) $(2 - \ln x)^{\frac{3}{2}}$ (l) $\dfrac{1}{e^x - e^{-x}}$

Derivatives of sec x, cosec x, cot x

We can use the method of the last section to work out two of the derivatives of the reciprocal trigonometric functions.

$$y = \sec x = \frac{1}{\cos x} = (\cos x)^{-1}$$

$$\Rightarrow \frac{dy}{dx} = -(\cos x)^{-2}(-\sin x) = \frac{\sin x}{\cos^2 x} = \frac{\sin x}{\cos x} \times \frac{1}{\cos x}$$

$$= \tan x \sec x$$

$$y = \csc x = \frac{1}{\sin x} = (\sin x)^{-1}$$

$$\Rightarrow \frac{dy}{dx} = -(\sin x)^{-2}(\cos x) = \frac{-\cos x}{\sin^2 x} = \frac{-\cos x}{\sin x} \times \frac{1}{\sin x}$$

$$= -\cot x \csc x$$

The remaining function is a quotient

$$y = \cot x = \frac{\cos x}{\sin x} \Rightarrow \frac{dy}{dx} = \frac{\sin x (-\sin x) - \cos x (\cos x)}{(\sin x)^2}$$

$$= -\frac{(\sin^2 x + \cos^2 x)}{\sin^2 x}$$

$$= \frac{-1}{\sin^2 x} = -\csc^2 x$$

If you can remember the derivatives of tan and sec, you can work out the derivatives of cotan (= cot) and cosec: just add co to everything but make it minus:

$$\tan x \to \sec^2 x \Rightarrow \cot(\text{an}) x \to -\csc^2 x$$
$$\sec x \to \sec x \tan x \Rightarrow \csc x \to -\csc x \cot x$$

Practice questions F

1 Write down the derivatives of

(a) $\cot 5x$ (b) $\csc 2x$ (c) $\sec \frac{x}{2}$ (d) $\cot^2 x$ (e) $(1 + \csc x)^4$ (f) $\frac{1}{1 + \sec x}$

2 Show by differentiating $\frac{\sin x}{\cos x}$ that the derivative of $\tan x$ is $\sec^2 x$.

Partial fractions and differentiation

In order to differentiate a function like:

$$y = \frac{5x - 7}{(x - 2)(x - 1)}$$

and especially if you want to find the higher order derivatives like $\frac{d^2y}{dx^2}$, $\frac{d^3y}{dx^3}$ etc, it is often easier to express the function in partial fractions first of all and then differentiate these parts separately. For the above function, for example:

$$y = \frac{5x - 7}{(x - 2)(x - 1)} \equiv \frac{A}{x - 2} + \frac{B}{x - 1} \equiv \frac{A(x - 1) + B(x - 2)}{(x - 2)(x - 1)}$$

i.e. $5x - 7 \equiv A(x - 1) + B(x - 2)$

Putting $x = 1$: $-2 = -B \Rightarrow B = 2$

$x = 2$: $3 = A \Rightarrow A = 3$

i.e. $y = \frac{3}{x - 2} + \frac{2}{x - 1}$

By expressing these fractions as powers of brackets, we can differentiate as many times s we like quite easily:

$$y = 3(x - 2)^{-1} + 2(x - 1)^{-1}$$

$$\frac{dy}{dx} = -3(x - 2)^{-2} - 2(x - 1)^{-2}$$

$$\frac{d^2y}{dx^2} = 6(x - 2)^{-3} + 4(x - 1)^{-3}$$

$$\frac{d^3y}{dx^3} = -18(x - 2)^{-4} - 12(x - 1)^{-4} \text{ etc.}$$

Practice questions G

1. If $f(x) = \dfrac{1}{(x-1)(2-x)}$, express $f(x)$ in partial fractions and hence find $f''(x)$

2. If $y = \dfrac{x^2 + 3x + 1}{(x+1)(x+2)}$, express y in the form $A + \dfrac{B}{x+1} + \dfrac{C}{x+2}$, where A, B and C are constants.

   Hence find the value of $\dfrac{dy}{dx}$ when $x = 0$.

3. Given that $f(x) = \dfrac{13x + 16}{(x-3)(3x+2)}$, express $f(x)$ in partial fractions and hence find the value of $f'(2)$.

Functions of ln x

When we have to differentiate an expression like $\ln [f(x)]$ we can very often simplify the problem by using the log laws.

Example Differentiate the following with respect to x, simplifying your answers as far as possible.

(a) $\ln \sqrt{1 - x^2}$
(b) $\ln \left(x \sqrt{1 - x^2} \right)$
(c) $\ln \left(\dfrac{1 + \sin x}{1 - \sin x} \right)$

Solution (a) Using $\log A^n = n \log A$,

If $y = \ln \sqrt{1 - x^2} = \ln (1 - x^2)^{\frac{1}{2}} = \dfrac{1}{2} \ln (1 - x^2) \Rightarrow \dfrac{dy}{dx} = \dfrac{1}{2} \left[\dfrac{-2x}{1 - x^2} \right]$

$\qquad = \dfrac{-x}{1 - x^2}$

(b) Using $\log AB = \log A + \log B$

If $y = \ln \left(x \sqrt{1 - x^2} \right) = \ln x + \ln \sqrt{1 - x^2} = \ln x + \dfrac{1}{2} \ln (1 - x^2)$

$\Rightarrow \dfrac{dy}{dx} = \dfrac{1}{x} + \dfrac{1}{2} \left(\dfrac{-2x}{1 - x^2} \right) = \dfrac{1}{x} - \dfrac{x}{1 - x^2} = \dfrac{1 - x^2 - x^2}{x(1 - x^2)}$

$\qquad = \dfrac{1 - 2x^2}{x(1 - x^2)}$

(c) Using $\log \dfrac{A}{B} = \log A - \log B$, the expression becomes

$\ln (1 + \sin x) - \ln (1 - \sin x)$. This differentiates to

$\dfrac{\cos x}{1 + \sin x} - \dfrac{(-\cos x)}{1 - \sin x} = \dfrac{\cos x (1 - \sin x) + \cos x (1 + \sin x)}{(1 + \sin x)(1 - \sin x)}$

$\qquad = \dfrac{\cos x - \cos x \sin x + \cos x + \cos x \sin x}{1 - \sin^2 x}$

$\qquad = \dfrac{2 \cos x}{\cos^2 x} = \dfrac{2}{\cos x} = 2 \sec x$

P3 Section 4

Practice questions H

1. Differentiate the following with respect to x, simplifying your answers where possible.

   (a) $\ln\left(\dfrac{1}{1-x^2}\right)$ (b) $\ln\sqrt{\left(\dfrac{1+x}{1-x}\right)}$ (c) $\ln\left(\dfrac{x}{2x-1}\right)$ (d) $\ln(\sin x \cos x)$ (e) $\ln\sqrt{\left(\dfrac{1-\cos x}{1+\cos x}\right)}$

Parametric differentiation

OCR P3 5.3.4 (d)

When x and y are expressed in the form $x = f(t)$ and $y = g(t)$, where f and g are functions of the parameter t, we find $\dfrac{dy}{dx}$ by finding first of all $\dfrac{dx}{dt}$ and $\dfrac{dy}{dt}$ and then putting

$$\dfrac{dy}{dx} = \dfrac{dy}{dt} \bigg/ \dfrac{dx}{dt}$$

Example

Find an expression for $\dfrac{dy}{dx}$ when

(a) $x = 5t^2 + 2$ and $y = t^3 - t$

(b) $x = \cos\theta$ and $y = \cos 2\theta$

Solution

(a) $\dfrac{dx}{dt} = 10t$ and $\dfrac{dy}{dt} = 3t^2 - 1$

Then $\dfrac{dy}{dx} = \dfrac{dy}{dt} \bigg/ \dfrac{dx}{dt} = \dfrac{3t^2 - 1}{10t}$

(b) $\dfrac{dx}{d\theta} = -\sin\theta$ and $\dfrac{dy}{d\theta} = -2\sin 2\theta$

Then $\dfrac{dy}{dx} = \dfrac{-2\sin 2\theta}{-\sin\theta} = \dfrac{2[2\sin\theta \cos\theta]}{\sin\theta} = 4\cos\theta$

The questions can involve other parts of the syllabus: the quotient rule for differentiation, for example

Example

If $x = \dfrac{t}{1+t^2}$ and $y = \dfrac{1}{1+t^2}$

show that $(1 - t^2)\dfrac{dy}{dx} + 2t = 0$

Solution

$\dfrac{dx}{dt} = \dfrac{(1+t^2)(1) - t(2t)}{(1+t^2)^2} = \dfrac{1+t^2-2t^2}{(1+t^2)^2} = \dfrac{1-t^2}{(1+t^2)^2}$

$y = (1+t^2)^{-1} \Rightarrow \dfrac{dy}{dt} = -(1+t^2)^{-2}(2t) = \dfrac{-2t}{(1+t^2)^2}$

$$\frac{dy}{dx} = \frac{dy}{dt} \bigg/ \frac{dx}{dt} = \frac{-2t}{(1+t^2)^2} \times \frac{(1+t^2)^2}{1-t^2} = \frac{-2t}{1-t^2}$$

$$\Rightarrow (1-t^2)\frac{dy}{dx} + 2t = (1-t^2) \times \frac{-2t}{1-t^2} + 2t = -2t + 2t = 0 \quad \text{as required}$$

When trigonometric functions are involved, solutions may have to be simplified by using one of the identities.

Example If $x = \sin 4\theta + 2 \sin 2\theta$ and $y = \cos 4\theta - 2 \cos 2\theta$ where $\frac{-\pi}{6} < \theta < \frac{\pi}{6}$, show that $\frac{dy}{dx} = -\tan\theta$.

Solution $\frac{dx}{d\theta} = 4\cos 4\theta + 4\cos 2\theta$, $\frac{dy}{d\theta} = -4\sin 4\theta + 4\sin 2\theta$

$$\frac{dy}{dx} = \frac{dy}{d\theta} \bigg/ \frac{dx}{d\theta} = \frac{-4\sin 4\theta + 4\sin 2\theta}{4\cos 4\theta + 4\cos 2\theta}$$

$$= \frac{\sin 2\theta - \sin 4\theta}{\cos 4\theta + \cos 2\theta}$$

using factor formulae $= \frac{2\cos 3\theta \sin(-\theta)}{2\cos 3\theta \cos\theta}$

$$= \frac{\sin(-\theta)}{\cos\theta} = \frac{-\sin\theta}{\cos\theta} = -\tan\theta$$

Practice questions I

1 Find an expression for $\frac{dy}{dx}$ when
 (a) $x = t^2$, $y = t^3$
 (b) $x = 2\sin\theta$, $y = 3\cos\theta$
 (c) $x = \sin 2\theta$, $y = \cos 4\theta$
 (d) $x = \frac{t}{1+t}$, $y = \frac{t^2}{1+t}$
 (e) $x = at$, $y = \frac{a}{t}$
 (f) $x = \frac{1}{(4-t)^2}$, $y = \frac{t}{4-t}$

2 If $x = 2t + \sin 2t$ and $y = \cos 2t$, where t is a parameter with $0 < t < \frac{\pi}{2}$, show that $\frac{dy}{dx} = -\tan t$.

3 If $x = 2t - \ln(2t)$ and $y = t^2 - \ln(t^2)$ where t is a positive parameter, find the value of t at the point on the curve at which the gradient is 2.

4 If $x = a\cos^3\theta$ and $y = a\sin^3\theta$, with $0 < \theta < \frac{\pi}{2}$, show that $\frac{dy}{dx} = -\tan\theta$.

5 If $x = a\theta - a\sin\theta$ and $y = a - a\cos\theta$, with $0 < \theta < \frac{\pi}{2}$, show that $\frac{dy}{dx} = \cot\left(\frac{1}{2}\theta\right)$.

6 A curve has parametric equations
$$x = t + \frac{1}{t}, \quad y = t + 1,$$
where $t \neq 0$. Express $\frac{dy}{dx}$ in terms of t, and hence find the values of t for which the gradient of the curve is 2.

7 The parametric equations of a curve C are
$$x = t + e^t, \quad y = t + e^{-t}$$
Find $\frac{dy}{dx}$ in terms of t, and hence find the coordinates of the stationary point of C.

8 The parametric equations of a curve are
$$x = e^{2t} - 5t, \quad y = e^{2t} - 2t$$
Find $\frac{dy}{dx}$ in terms of t.

Find the exact value of t at the point on the curve where the gradient is 2.

Tangents and normals

OCR P1 5.1.5 (d)

To find the equation of a tangent or a normal when y is an implicit function of x, or when x and y are given in terms of a parameter, follows the same pattern that we looked at in the last unit: find $\dfrac{dy}{dx}$, the value of this and the x- and y-coordinates, and the corresponding equation. Here are some examples of these.

Example

The equation of a curve is $x^2 + 4xy + y^2 = 25$. Find the equation of the tangent to the curve at the point where the curve meets the positive x-axis.

Solution

We have to find the point first of all. Since it crosses the x-axis, $y = 0 \Rightarrow x^2 = 25$ $\Rightarrow x = 5$ since positive.

To find $\dfrac{dy}{dx}$ we differentiate the equation with respect to x:

$$2x + 4\left[y + x\dfrac{dy}{dx}\right] + 2y\dfrac{dy}{dx} = 0$$

When $x = 5$ and $y = 0$,

$$10 + 4\left[0 + 5\dfrac{dy}{dx}\right] + 0 \times \dfrac{dy}{dx} = 0$$

$$\Rightarrow \quad 10 + 20\dfrac{dy}{dx} = 0 \Rightarrow \dfrac{dy}{dx} = -\dfrac{1}{2}$$

Equation of the tangent is $y - 0 = -\dfrac{1}{2}(x - 5)$

$$\Rightarrow \quad 2y = -x + 5$$

or $\quad x + 2y = 5$

Example

The parametric equations of a curve are

$$x = 2\theta - \sin 2\theta$$
$$y = 1 - \cos 2\theta$$

The tangent and the normal to the curve P where $\theta = \dfrac{\pi}{4}$ meet the y-axis at L and M respectively. Show that the area of triangle PLM is $\dfrac{1}{4}(\pi - 2)^2$.

Solution

$$\dfrac{dx}{d\theta} = 2 - 2\cos 2\theta, \quad \dfrac{dy}{d\theta} = 2\sin 2\theta$$

$$\dfrac{dy}{dx} = \dfrac{dy}{d\theta} \bigg/ \dfrac{dx}{d\theta} = \dfrac{2\sin 2\theta}{2 - 2\cos 2\theta}$$

We can simplify this by substituting $\sin 2\theta = 2\sin\theta\cos\theta$, $2 - 2\cos 2\theta = 2(2\sin^2\theta)$ to give

$$\dfrac{dy}{dx} = \dfrac{2(2\sin\theta\cos\theta)}{2(2\sin^2\theta)} = \dfrac{\cos\theta}{\sin\theta} = \cot\theta$$

when $\theta = \dfrac{\pi}{4}$, $\dfrac{dy}{dx} = \cot\theta = 1$, $\quad x = \left[\dfrac{\pi}{2} - \sin\dfrac{\pi}{2}\right] = \left[\dfrac{\pi}{2} - 1\right]$

and $y \quad = \left[1 - \cos\dfrac{\pi}{2}\right] = 1$

Tangent has equation $y - 1 = 1\left(x - \left(\frac{\pi}{2} - 1\right)\right)$

$$y - 1 = x - \frac{\pi}{2} + 1$$

When $x = 0$ (on the y-axis) $y = 2 - \frac{\pi}{2}$ (Point L)

Normal has equation $y - 1 = -1\left(x - \left(\frac{\pi}{2} - 1\right)\right)$

$$y - 1 = -x + \frac{\pi}{2} - 1$$

When $x = 0$, $y = \frac{\pi}{2}$ (Point M)

Figure 4.1

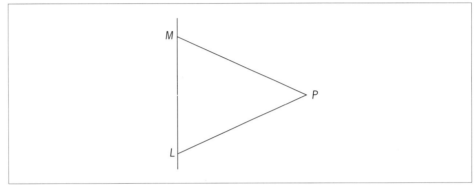

Area Δ is $\frac{1}{2}$ base × height = $\frac{1}{2}$ ML × x-coordinate of P

$ML = \frac{\pi}{2} - \left(2 - \frac{\pi}{2}\right) = \pi - 2$

x-coordinate of P is $\frac{\pi}{2} - 1$ as before

$\Rightarrow$ Area is $\frac{1}{2} \times (\pi - 2) \times \left(\frac{\pi}{2} - 1\right)$

$= \frac{1}{2} \times (\pi - 2) \times \frac{1}{2}(\pi - 2)$

$= \frac{1}{4}(\pi - 2)^2$ as required.

Practice questions J

1. Find the equation of the normal at the origin to the curve with equation

   $y - x + 3e^{-y} = 3$

2. If $x = t^2 - 4$ and $y = 3t^4 + 8t^3$ where t is a parameter, find the equation of the tangent to the curve at the point where $t = -1$, giving your answer in the form $ax + by + c = 0$, where a, b and c are integers.

3. Find the equation of the normal to the curve $3y^2 - x^2 = 3$ at the point $(3, 2)$

4. The parametric equations of a curve are

   $x = \dfrac{1+t}{1-2t}$ and $y = \ln(t-1)$,

   where $t > 1$. Find the value of $\dfrac{dy}{dx}$ at the point where $t = 2$.

   Hence find the equation of the tangent to the curve at this point.

5 The parametric equations of a curve are
$$x = 2t^2, \ y = 4t,$$
where the parameter t takes all real values. Show that the tangent to the curve at the point $P(2p^2, 4p)$ has equation
$$py = x + 2p^2$$

6 The coordinates (x, y) of a point on a curve are given in terms of a parameter t by
$$x = t^2 + 1, \ y = t^3 - t.$$
(a) Find the equation of the tangent to the curve at the point where $t = 1$,
(b) Find the values of t for which both x and y have the same rate of change with respect to t.
[The rate of change of a variable p with respect to t is $\dfrac{dp}{dt}$.]

7 A curve is defined by the parametric equations
$$x = t^2, \ y = \frac{2}{t} \ (t \neq 0)$$
Show that an equation of the tangent to the curve at the point $P\left(p^2, \dfrac{2}{p}\right)$ is
$$x + p^3 y = 3p^2$$
Given that this tangent passes through the point $(1, 2)$, determine the possible values of p.

8 Find the equation of the normal to the hyperbola $xy = c^2$ at the point $P\left(cp, \dfrac{c}{p}\right)$.
The normal cuts the curve again at $Q\left(cq, \dfrac{c}{q}\right)$.
Show that $p^3 q + 1 = 0$.

Practice questions K (Miscellaneous and harder)

1 If $y = \ln(1 + \sin x)$, find $\dfrac{dy}{dx}$ and $\dfrac{d^2y}{dx^2}$ in terms of x and show that $\cos x \dfrac{d^2y}{dx^2} + \dfrac{dy}{dx} = 0$

2 If $y = (3 + 4x)e^{-2x}$, find $\dfrac{dy}{dx}$ and $\dfrac{d^2y}{dx^2}$ and show that $\dfrac{d^2y}{dx^2} + 4\dfrac{dy}{dx} + 4y = 0$

3 If $y = \dfrac{1 - \sin x}{\cos x}$, show that $\dfrac{dy}{dx} = -\dfrac{1}{1 + \sin x}$
and that $\dfrac{d^2y}{dx^2} = \dfrac{\cos x}{(1 + \sin x)^2}$

4 If $y = \sqrt{\cos x}$, show that
$$y \dfrac{d^2y}{dx^2} + \left(\dfrac{dy}{dx}\right)^2 + \dfrac{1}{2} y^2 = 0$$

5 Given that $y = x - \ln(\sec x + \tan x)$, find $\dfrac{dy}{dx}$, giving your answer in a form that involves only one trigonometric function.

6 Given that $x + y = (x - y)^2$
Show that $\dfrac{dy}{dx} = \dfrac{2x - 2y - 1}{2x - 2y + 1}$

7 Given that $y = \ln(1 + \sin x)$, show that
$$\dfrac{d^2y}{dx^2} + e^{-y} = 0$$

8 If $y = \sqrt{5x^2 + 3}$, show that
$$y \dfrac{d^2y}{dx^2} + \left(\dfrac{dy}{dx}\right)^2 = 5$$

9 If $x = 5a \sec \theta$ and $y = 3a \tan \theta$, where $\dfrac{-\pi}{2} < \theta < \dfrac{\pi}{2}$, find the coordinates of the point on the curve at which the normal is parallel to the line $y = x$.

10 If $x = e^\theta \cos \theta$ and $y = e^\theta \sin \theta$, show that
$$\dfrac{dy}{dx} = \tan\left(\theta + \dfrac{\pi}{4}\right)$$

11 A curve has parametric equations
$$x = \sec t + \tan t, \ y = \csc t + \cot t, \ 0 < t < \dfrac{\pi}{2}.$$
Show that $\dfrac{dy}{dx} = -\dfrac{1 - \sin t}{1 - \cos t}$

12 If $x = \cos 2\theta + 2\cos \theta, \ y = \sin 2\theta - 2\sin \theta$,
show that $\dfrac{dy}{dx} = \tan \dfrac{\theta}{2}$.

Find the equation of the normal to the curve at the point where $\theta = \dfrac{\pi}{2}$

13 Given that $f(x) = \dfrac{\sin x}{2 - \cos x}$
find the greatest and least values of $f(x)$.

14 Given that $y^2 - 5xy + 8x^2 = 2$,

show that $\dfrac{dy}{dx} = \dfrac{5y - 16x}{2y - 5x}$

The distinct points P and Q on the curve $y^2 - 5xy + 8x^2 = 2$, each have x-coordinates 1. The normals to the curve at P and Q meet at the point N. Calculate the coordinates of N.

15 Find the four values of x for which the curve $y = \sqrt{2}\,e^{-x}\cos x$ has turning points in the range $0 \leq x \leq 4\pi$, writing your values in ascending order. Show that the corresponding y-values form the first four terms of a geometric progression.

16 A curve is given by the parametric equations

$$x = \tfrac{1}{2}\left(t + \tfrac{1}{t}\right), \quad y = \tfrac{1}{2}\left(t - \tfrac{1}{t}\right), \quad y \neq 0$$

(a) Express $x + y$ and $x - y$ in terms of t and hence find the cartesian equation of the curve.

(b) Find $\dfrac{dy}{dx}$ in terms of t and hence find in its simplest form the gradient of the tangent to the curve at the point P with parameter p.

(c) Show that the equation of the tangent at P is
$$(p^2 + 1)x - (p^2 - 1)y = 2p$$

SUMMARY EXERCISE

1 Given that $y = \dfrac{1 - x}{1 + x}$, show that

$$\dfrac{d^2y}{dx^2} = \dfrac{4}{(1 + x)^3}$$

2 Given that $y = \sin(x^3)$, find $\dfrac{d^2y}{dx^2}$.

3 Given that $e^{2x} + e^{2y} = xy$, find $\dfrac{dy}{dx}$ in terms of x and y.

4 The parametric equations of a curve are

$$x = 2\cos t, \quad y = 5 + 3\cos 2t,$$

where $0 < t < \pi$. Express $\dfrac{dy}{dx}$ in terms of t, simplifying your answer, and hence show that the gradient at any point on the curve is less than 6.

5 A curve C is given by the equation

$$y^3 + y^2 + y = x^2 - 2x$$

(a) Show that the point $(3, 1)$ is the only point of intersection of the line $x = 3$ and the curve.

(b) Show that the tangent to C at the point $(-1, 1)$ has equation $2x + 3y - 1 = 0$.

6 $f(x) \equiv \dfrac{x}{x^2 + 2}$, $x \in \mathbb{R}$

Find the set of values of x for which $f'(x) < 0$.

7 Differentiate, with respect to x,

(a) $\dfrac{\sin x}{x}$, $x > 0$,

(b) $\ln\left(\dfrac{1}{x^2 + 9}\right)$

Given that $y = x^x$, $x > 0$, $y > 0$, by taking logarithms

(c) show that $\dfrac{dy}{dx} = x^x(1 + \ln x)$

8 The curve C has parametric equations

$$x = t^3, \quad y = t^2, \quad t > 0$$

(a) Find an equation of the tangent to C at A $(1, 1)$

Given that the line l with equation $3y - 2x + 4 = 0$ cuts the curve C at point B:

(b) find the coordinates of B

(c) prove that the line l only cuts C at the point B.

9 $f(x) \equiv e^{2x}\sin 2x$, $0 \leq x \leq \pi$.

(a) Find the values of x for which $f(x) = 0$ giving your answers in terms of π.

(b) Use calculus to find the coordinates of the turning points on the graph of

$$y = f(x)$$

(c) Show that $f''(x) = 8e^{2x}\cos 2x$

(d) Hence, or otherwise, determine which turning point is a maximum and which is a minimum.

10 The parametric equations of a curve are
$$x = t - e^{2t}, \quad y = t + e^{2t},$$
where t takes all real values.

(a) Express $\dfrac{dy}{dx}$ in terms of t.

(b) Hence find the value of t for which the gradient of the curve is 3, giving your answer in the form $a \ln b$.

11 The parametric equations of a curve are
$$x = \cos^3 t, \quad y = \sin^3 t$$
where $0 < t \le \tfrac{1}{2}\pi$.

Find and simplify an expression for the gradient of the curve at the point with parameter t.

Find the equation of the tangent at the point with coordinates $\left(\tfrac{3}{8}\sqrt{3}, \tfrac{1}{8}\right)$, giving your answer in the form $y = mx + c$, where the exact values of m and c should be stated.

12 Use differentiation to find the x-coordinates of the points at which the graph of $y = \dfrac{x^2}{2x - 1}$ has either a maximum or a minimum and distinguish between them.

SUMMARY

When you have finished this section, you should:

- know the derivative of $\sin x$, $\cos x$ and $\tan x$

- know how to differentiate products of functions

- know how to differentiate quotients

- know the derivative of a^x and how to derive it

- be familiar with the method of finding the derivatives of the composition of simple functions such as $\cos(x^2)$

- know that $\dfrac{dy}{dx} = 1 \Big/ \dfrac{dx}{dy}$

- be able to find the derivative of powers of functions

- know that the derivative of a function involving ln can frequently be simplified using the log laws

- be able to work out the derivates of the reciprocal trigonometric functions: $\sec x$, $\operatorname{cosec} x$ and $\cot x$

- know how to use partial fractions in order to differentiate a rational expression

- know how to find a function for the gradient when x and y are expressed in terms of a parameter

- know how to find the equation of a tangent or normal to a curve whose equation is expressed in implicit or parametric form

ANSWERS

Practice questions A

1. (a) $3x^2 \ln x + x^2$ (b) $xe^x(x+2)$
   (c) $\cos^2 - \sin^2 x = \cos 2x$
   (d) $e^x(\cos x - \sin x)$ (e) $\tan x + x \sec^2 x$
   (f) $\cos x \ln x + \dfrac{\sin x}{x}$
   (g) $x^3(x \cos x + 4 \sin x)$
   (h) $3\left[\dfrac{\cos x}{x} - \sin x \ln 5x\right]$

Practice questions B

1. (a) $\dfrac{x(2-x)}{(1-x)^2}$ (b) $\dfrac{1 + \sin x + \cos x}{(1 + \cos x)^2}$
   (c) $\dfrac{1 - 2 \ln x}{x^3}$ (d) $\dfrac{\sec^2 x - \tan x}{e^x}$
   (e) $\dfrac{1 - x^2}{(1 + x^2)^2}$ (f) $\dfrac{2e^x}{(1 - e^x)^2}$
   (g) $\dfrac{e^x(x-2)}{x^3}$ (h) $\sec^2 x$

2. $-e^{-3}$

3. $(0, 0), \left(1, \dfrac{1}{2}\right), \left(-1, \dfrac{1}{2}\right)$

5. $\dfrac{6x}{(2x^2 + 1)^2}, \ x > 0$

Practice questions C

1. (a) $3y^2 \dfrac{dy}{dx}$ (b) $\cos y \dfrac{dy}{dx}$
   (c) $\dfrac{1}{y} \dfrac{dy}{dx}$ (d) $e^y \dfrac{dy}{dx}$
   (e) $3 \sec^2 y \dfrac{dy}{dx}$ (f) $\sin y \dfrac{dy}{dx}$

2. (a) $2xy^2 + 2x^2y \dfrac{dy}{dx}$
   (b) $e^y + xe^y \dfrac{dy}{dx}$
   (c) $3x^2 \cos y - x^3 \sin y \dfrac{dy}{dx}$
   (d) $\dfrac{dy}{dx} \sin x + y \cos x$
   (e) $4y^3 + 12xy^2 \dfrac{dy}{dx}$
   (f) $\sec^2 x \tan y + \tan x \sec^2 y \dfrac{dy}{dx}$
   (g) $\left(1 + \dfrac{dy}{dx}\right) y^2 + 2(x + y) y \dfrac{dy}{dx}$
   (h) $\left(1 + \dfrac{dy}{dx}\right) e^{x+y}$

3. (a) $-\dfrac{x^2}{y^2}$ (b) $\dfrac{2x - 1}{2y}$ (c) $\dfrac{\cos x}{2y + 1}$
   (d) $\dfrac{3xy - 2y^2}{6xy - 3x^2}$ (e) $\dfrac{4x^3 - y}{x + 2y}$
   (f) $\dfrac{\sec^2 x - \cos x}{\cos y}$

4. -1

5. $-\dfrac{1}{4}$

6. $-\dfrac{2}{5}$

7. $\dfrac{y^2}{x(1 - y \ln x)}$

8. $\dfrac{x(2y^2 - x)}{2y(y - x^2)}, \ \left(2^{\frac{1}{3}}, 2^{-\frac{1}{3}}\right)$

9. (a) $5^x (\ln 5)$ (b) $\left(\dfrac{1}{2}\right)^x \left(\ln \dfrac{1}{2}\right)$

Practice questions D

1. (a) $e^x \cos(e^x)$ (b) $-4e^{-4x}$
   (c) $\dfrac{4}{1 + 4x}$ (d) $3x^2 \sec^2(x^3)$
   (e) $-3 \sin 3x$ (f) $-\sin x \, e^{\cos x}$
   (g) $\dfrac{\cos}{1 + \sin x}$ (h) $\dfrac{1}{x} \cos(\ln x)$
   (i) $\sin\left(\dfrac{\pi}{2} - x\right)$ (j) $3 \sec^2\left(3x + \dfrac{\pi}{4}\right)$

Practice questions E

1. (a) $20(1 + 4x)^4$ (b) $-6e^x(3 - e^x)^5$
   (c) $\dfrac{-2}{(1 + 2x)^2}$ (d) $\dfrac{x}{(1 - x^2)^{\frac{3}{2}}}$
   (e) $-4 \cos^3 x \sin x$ (f) $\dfrac{\cos x}{2\sqrt{\sin x}}$
   (g) $5 \sec^2 x (1 + \tan x)^4$
   (h) $4(\sin x + \cos x)^3 (\cos x - \sin x)$
   (i) $6 \tan 3x \sec^2 3x$
   (j) $\dfrac{-6x^2}{\sqrt{1 - 4x^3}}$
   (k) $\dfrac{-3 \sqrt{2 - \ln x}}{2x}$
   (l) $\dfrac{-(e^x + e^{-x})}{(e^x - e^{-x})^2}$

Practice questions F

1. (a) $-5 \csc^2 5x$ (b) $-2 \csc 2x \cot 2x$
   (c) $\frac{1}{2} \sec \frac{x}{2} \tan \frac{x}{2}$ (d) $-2 \cot x \csc^2 x$
   (e) $-4(1 + \csc x)^3 \csc x \cot x$
   (f) $\dfrac{-\sec x \tan x}{(1 + \sec x)^2}$

Practice questions G

1. $\dfrac{2}{(x-1)^3} + \dfrac{2}{(2-x)^3}$
2. $1 - \dfrac{1}{x+1} + \dfrac{1}{x+2} ; \dfrac{3}{4}$
3. $\dfrac{5}{x-3} - \dfrac{2}{3x+2} ; \dfrac{-157}{32}$

Practice questions H

1. (a) $\dfrac{2x}{1-x^2}$ (b) $\dfrac{1}{1-x^2}$ (c) $\dfrac{-1}{x(2x-1)}$
   (d) $\dfrac{\cos 2x}{\frac{1}{2}\sin 2x} = 2\cot 2x$ (e) $\csc x$

Practice questions I

1. (a) $\dfrac{3t}{2}$ (b) $\dfrac{-3}{2}\tan\theta$
   (c) $\dfrac{-4\sin 4\theta}{\cos 2\theta} = -8\sin 2\theta$ (d) $2t + t^2$
   (e) $\dfrac{-1}{t^2}$ (f) $2(4-t)$

3. 2

6. $\dfrac{t^2}{t^2-1}$, $\pm\sqrt{2}$

7. $\dfrac{1-e^t}{1+e^t}$; $(1, 1)$

8. $\dfrac{2e^{2t}-2}{2e^{2t}-5}$; $\ln 2$

Practice questions J

1. $y = 2x$ 2. $6x + y + 23 = 0$
3. $2x + y - 8 = 0$ 4. $3, y = 3x + 3$
6. (a) $y = x - 2$ (b) $\dfrac{-1}{3}, 1$
7. $1, \dfrac{-1}{2}$
8. $py = p^3 x + c - cp^4$

Practice questions K

1. $\dfrac{dy}{dx} = \dfrac{\cos x}{1+\sin x}$, $\dfrac{d^2y}{dx^2} = \dfrac{-1}{1+\sin x}$
2. $\dfrac{dy}{dx} = -2e^{-2x}(1+4x)$, $\dfrac{d^2y}{dx^2} = 4e^{-2x}(4x-1)$
5. $1 - \sec x$
9. $\left(\dfrac{25a}{4}, \dfrac{-9a}{4}\right)$
12. $x + y + 3 = 0$
13. $\dfrac{1}{\sqrt{3}}, -\dfrac{1}{\sqrt{3}}$
14. $\left(\dfrac{1}{7}, \dfrac{15}{7}\right)$
15. $\dfrac{3\pi}{4}, \dfrac{7\pi}{4}, \dfrac{11\pi}{4}, \dfrac{15\pi}{4}$
16. (a) $x + y = t$, $x - y = \dfrac{1}{t} \Rightarrow x^2 - y^2 = 1$
    (b) $\dfrac{t^2+1}{t^2-1}, \dfrac{p^2+1}{p^2-1}$

SECTION 5
Integration

INTRODUCTION This is another major topic in this unit. From the work on differentiation in the previous section we can derive a whole new range of integrals of standard functions. Very much work has been done in the area of integration and many techniques has been invented to cope with different types of integrals which arise in the solution of problems, practical or otherwise. We look at two of these, substitution and integration by parts, both of which are very powerful methods with broad applications. We further extend our work with area and volumes by looking at the area under a curve whose equation is given in parametric form.

Standard integrals

OCR P3 5.3.5 (a)

We can reverse the differentiation from the last section to give three standard integrals:

$$\int \sin x \, dx = -\cos x + C$$
$$\int \cos x \, dx = \sin x + C$$
$$\int \sec^2 x \, dx = \tan x + C$$

If we differentiate $\sin 4x$ we end up with $4\cos 4x$. This means that integrating $4\cos 4x$ would give $\sin 4x$,

$$\int 4\cos 4x \, dx = \sin 4x + C$$
$$\int \cos 4x \, dx = \tfrac{1}{4}\sin 4x + C$$

Similarly $\int \cos 7x \, dx = \tfrac{1}{7}\sin 7x + C$ and in general $\int \cos ax \, dx = \tfrac{1}{a}\sin ax + C$. Each of the three standard integrals has an equivalent form, plus we can add the corresponding result for e^{ax}.

$$\int \sin ax \, dx = -\tfrac{1}{a}\cos ax + C$$
$$\int \cos ax \, dx = \tfrac{1}{a}\sin ax + C$$
$$\int \sec^2 ax \, dx = \tfrac{1}{a}\tan ax + C$$
$$\int e^{ax} \, dx = \tfrac{1}{a}e^{ax} + C$$

Practice questions A

1 Write down the integral of the following functions:
(a) $\sin 2x$
(b) $\cos 3x$
(c) $\sec^2 5x$
(d) e^{3x}
(e) $2\cos 4x$
(f) $-3\sin 3x$
(g) $2\sec^2 \tfrac{x}{3}$
(h) $\dfrac{1}{e^{4x}}$
(i) $\sin\left(\tfrac{\pi}{6} - x\right)$
(j) $2\cos\left(3x - \tfrac{\pi}{4}\right)$
(k) $e^{3-\tfrac{1}{2}x}$
(l) $4\sin \tfrac{x}{2}$

Recognition

OCR P3 5.3.5 (d)

One important group of integrals comes from reversing the derivative of a log function. Suppose for example that we differentiated $y = \ln(1 + x^2)$

$$y = \ln(1 + x^2) \Rightarrow \frac{dy}{dx} = \frac{2x}{1 + x^2}$$

Reversing this,

$$\int \frac{2x}{1 + x^2}\, dx = \ln(1 + x^2) + C$$

So whenever we have a derivative over its corresponding function, or nearly, give or take a constant, we see whether we can apply the formula

$$\int \frac{\Box}{\Box} = \ln \Box + C$$

This is particularly important for denominators with linear functions, as we shall see when we come to integrating partial fractions. There is a formula in this case,

$$\int \frac{1}{ax + b}\, dx = \frac{1}{a} \ln(ax + b) + C$$

Note that strictly speaking, we should include the modulus sign, $\ln|ax + b|$, rather than $\ln(ax + b)$ – see pp. 93–4 of P2 for the reason. When no limits are involved, it is acceptable to use just brackets, so long as you remember to use the modulus signs for definite integrals. Some examples of this formula are:

$$\int \frac{1}{2x + 1}\, dx = \frac{1}{2} \ln(2x + 1) + C$$

$$\int \frac{1}{3 - 4x}\, dx = -\frac{1}{4} \ln(3 - 4x) + C$$

$$\int \frac{5}{7 - 3x}\, dx = -\frac{5}{3} \ln(7 - 3x) + C$$

When the denominator is not linear we check that the numerator is more or less the derivative of the denominator, apart from a constant.

Some examples of this:

$$\int \frac{3x^2}{1 + x^3}\, dx = \ln(1 + x^3) + C$$

$$\int \frac{e^x}{1 + e^x}\, dx = \ln(1 + e^x) + C$$

$$\int \frac{3x^2 + 1}{x^3 + x}\, dx = \ln(x^3 + x) + C$$

$$\int \frac{-2 \sec^2 x}{1 - 2 \tan x}\, dx = \ln(1 - 2 \tan x) + C$$

We may need to adjust the constant:

$$\int \frac{x^2}{1 + x^3}\, dx = \frac{1}{3} \ln(1 + x^3) + C$$

$$\int \frac{3x^4}{1 + x^5} = \frac{3}{5} \ln(1 + x^5) + C, \text{ etc.}$$

Note that we can integrate tan x and cot x using this method

$$\int \tan x \, dx = \int \frac{\sin x}{\cos x} \, dx = -\int \frac{-\sin x}{\cos x} \, dx$$
$$= -\ln \cos x + C$$
$$= \ln \frac{1}{\cos x} + C = \ln \sec x + C$$

We had to supply a negative sign here because the derivative of cos x is −sin x. We don't need this to integrate cot x.

$$\int \cot x \, dx = \int \frac{\cos x}{\sin x} \, dx = \ln \sin x + C$$

If you can't see what constant is required, you can always differentiate what you think is approximately the answer and adjust accordingly.

e.g. $\int \frac{2x}{1-4x^2} \, dx$

It's more or less the derivative on top, so we know it will involve $\ln(1-4x^2)$. Differentiating this would give $\frac{-8x}{1-4x^2}$ so we have to divide the −8x by −4 to give us the 2x we need.

$$\int \frac{2x}{1-4x^2} \, dx = -\frac{1}{4} \ln(1-4x^2) + C$$

Practice questions B

1 Find the following integrals:

(a) $\int \frac{1}{1+x} \, dx$ (b) $\int \frac{1}{1+3x} \, dx$ (c) $\int \frac{2}{x+4} \, dx$ (d) $\int \frac{3}{1+2x} \, dx$

(e) $\int \frac{2}{1-4x} \, dx$ (f) $\int \frac{5}{3-10x} \, dx$ (g) $\int \frac{7}{3x+2} \, dx$ (h) $\int \frac{6}{2+9x} \, dx$

2 Find:

(a) $\int \frac{x^3}{1+x^4} \, dx$ (b) $\int \frac{x^2}{1-2x^3} \, dx$ (c) $\int \frac{\sin 2x}{\cos 2x} \, dx$ (d) $\int \cot 3x \, dx$

(e) $\int \frac{x}{1-x^2} \, dx$ (f) $\int \frac{3x}{1+2x^2} \, dx$ (g) $\int \frac{2e^x}{1-e^x} \, dx$ (h) $\int \frac{2x-3}{x^2-3x+9} \, dx$

(i) $\int \frac{\sec^2 x}{1+\tan x} \, dx$ (j) $\int \frac{4x+1}{(2x-5)(x+3)} \, dx$

(k) $\int \frac{1}{x \ln x} \, dx \left(\frac{\frac{1}{x}}{\ln x} \right)$ (l) $\int \frac{1}{\sqrt{x}(1+\sqrt{x})} \, dx$

Powers of linear functions

This again comes from reversing an operation of differentiation, this time looking at a power of a linear function, something like $y = (ax + b)^n$. This gives

$$\frac{dy}{dx} = n(ax + b)^{n-1} \times a = na(ax + b)^{n-1}$$

Reversing this gives the formula

$$\int (ax + b)^n \, dx = \frac{(ax + b)^{n+1}}{a(n+1)} + C, \, n \neq -1$$

(If $n = -1$ we have a ln, as we saw in the last part.)

Some examples of this

$$\int (2x - 1)^5 \, dx = \frac{1}{12}(2x - 1)^6 + C$$

$$\int \frac{1}{\sqrt{3x - 1}} \, dx = \int (3x - 1)^{-\frac{1}{2}} \, dx = \frac{2}{3}(3x - 1)^{\frac{1}{2}} + C$$

$$\int \frac{1}{(1 - 4x)^2} \, dx = \int (1 - 4x)^{-2} \, dx = \frac{(1 - 4x)^{-1}}{4} + C$$

Practice questions C

1 Find

(a) $\int (3x + 2)^4 \, dx$
(b) $\int \frac{1}{(1 + x)^2} \, dx$
(c) $\int \sqrt{1 + 4x} \, dx$
(d) $\int \frac{4}{(1 - 3x)^3} \, dx$
(e) $\int \frac{1}{2(3 - x)^2} \, dx$
(f) $\int \frac{4}{\sqrt[3]{3 - 2x}} \, dx$

Integrals with partial fractions

OCR P3 5.3.5 (c)

Once we have found the partial fractions of a rational function we can then integrate much more easily. There are three types of fractions that could be involved and you need to recognise these and apply the appropriate formula – the last two are particular cases of the powers of linear functions that we have just looked at.

The types are

(a) $\int \frac{1}{ax + b} \, dx = \frac{1}{a} \ln | ax + b | + C$

(b) $\int \frac{1}{(ax + b)^2} \, dx = -\frac{1}{a(ax + b)} + C$

(c) $\int \frac{x}{ax^2 + b} \, dx = \frac{1}{2a} \ln | ax^2 + b | + C$

Here are three examples which cover these.

Example

Show that: $\int_1^2 \dfrac{6x+7}{(2x-1)(x+2)}\,dx = \ln 12$

Solution

We have a look first of all at the highest powers occurring on the top and bottom of the fraction – that on the top is less than that on the bottom in this case, so we don't need a preliminary division.

Suppose $\dfrac{6x+7}{(2x-1)(x+2)} \equiv \dfrac{A}{2x-1} + \dfrac{B}{x+2}$

$\equiv \dfrac{A(x+2) + B(2x-1)}{(2x-1)(x+2)}$

i.e. $6x + 7 \equiv A(x+2) + B(2x-1)$

$x = -2:\ -5 = -5B \quad \Rightarrow \quad B = 1$

$x = \dfrac{1}{2}:\ 10 = \dfrac{5A}{2} \quad \Rightarrow \quad A = 4$

Then $\int_1^2 \dfrac{6x+7}{(2x-1)(x+2)}\,dx = \int_1^2 \left(\dfrac{4}{(2x-1)} + \dfrac{1}{x+2}\right) dx$

$= \Big[2\ln(2x-1) + \ln(x+2)\Big]_1^2$

$= (2\ln 3 + \ln 4) - (2\ln 1 + \ln 3)$

$= \ln 3 + \ln 4$

$= \ln 12$

Example

Find $\int \dfrac{x^2 - 11}{(x+2)^2 (3x-1)}\,dx$

Solution

A repeated factor, so $\dfrac{A}{x+2} + \dfrac{B}{(x+2)^2} + \dfrac{C}{3x-1}$

i.e. $x^2 - 11 \equiv A(x+2)(3x-1) + B(3x-1) + C(x+2)^2$

$x = -2 \quad -7 = -7B \quad \Rightarrow B = 1$

$x = \dfrac{1}{2} \quad \dfrac{-98}{9} = \dfrac{49}{9} C \quad \Rightarrow C = -2$

x^2 coeff $\quad 1 = 3A + C \Rightarrow A = 1$

The integral is then

$\int \left(\dfrac{1}{x+2} + \dfrac{1}{(x+2)^2} - \dfrac{2}{3x-1}\right) dx$

$= \ln|x+2| - \dfrac{1}{x+2} - \dfrac{2}{3}\ln|3x-1| + C$

Example

Find:

$\int \dfrac{1 + 4x - x^2}{(x-2)(x^2+1)}\,dx$

Solution

Quadratic, $\dfrac{A}{x-2} + \dfrac{Bx+C}{x^2+1}$

i.e. $\quad 1 + 4x - x^2 \equiv A(x^2+1) + (Bx+C)(x-2)$

$x = 2 \qquad\qquad 5 \;= 5A \qquad \Rightarrow A = 1$

x^2 coeff $\qquad\quad -1 = A + B \qquad \Rightarrow B = -2$

constant $\qquad\quad\; 1 = A - 2C \qquad \Rightarrow C = 0$

We now have $\displaystyle\int\left(\dfrac{1}{x-2} - \dfrac{2x}{x^2+1}\right)dx$

$= \ln|x-2| - \ln|x^2+1| + C$

$= \ln\left|\dfrac{x-2}{x^2+1}\right| + C$

Practice questions D

1 Find the following

(a) $\displaystyle\int \dfrac{4}{(x-3)(x+1)}\,dx$
(b) $\displaystyle\int \dfrac{2}{x(x+1)(x-1)}\,dx$
(c) $\displaystyle\int \dfrac{x^2+3x+1}{(x+1)(x+2)}\,dx$

(d) $\displaystyle\int \dfrac{x-4}{(2x+1)(x-1)^2}\,dx$
(e) $\displaystyle\int \dfrac{1+12x-2x^2}{(x-3)(2x^2+1)}\,dx$
(f) $\displaystyle\int \dfrac{4}{x^2-4}\,dx$

2 Evaluate

(a) $\displaystyle\int_2^3 \dfrac{2x^2+1}{(x+2)(x-1)^2}\,dx$
(b) $\displaystyle\int_0^1 \dfrac{dx}{(x+1)(x+2)^2}$

(c) $\displaystyle\int_4^{12} \dfrac{3x+1}{2x^2+x}\,dx$
(d) $\displaystyle\int_1^2 \dfrac{x-3}{(x^2+3)(x+1)}\,dx$

3 Show the following

(a) $\displaystyle\int_7^{10} \dfrac{7}{x^2-5x-6}\,dx = \ln\left(\dfrac{32}{11}\right)$
(b) $\displaystyle\int_0^2 \dfrac{1}{(3-x)(1+x)}\,dx = \dfrac{1}{2}\ln 3$

(c) $\displaystyle\int_0^1 \dfrac{6-2x}{(x+1)(x^2+3)}\,dx = \ln 3$
(d) $\displaystyle\int_2^5 \dfrac{1}{2x^2-5x+3}\,dx = \ln\left(\dfrac{7}{4}\right)$

Substitution

OCR P3 5.3.5 (f)

It's not often in exams that we're presented with an integral that can be done straight away, with no preparatory work: usually they require a changing round of some description. The method of substitution that we study here is one that will enable us to solve quite a few types of otherwise difficult integrals.

Changing the variable

We can sometimes transform an integral containing an awkward function into something which is easier to handle by changing the variable. Let's compare, for example, the two integrals:

$$\int \dfrac{x}{\sqrt{x-1}}\,dx \qquad \text{and} \qquad \int \dfrac{x+1}{\sqrt{x}}\,dx$$

We can integrate the one on the right by separating and dividing:

$$\int \frac{x+1}{\sqrt{x}}\, dx = \int \left(\frac{x}{\sqrt{x}} + \frac{1}{\sqrt{x}}\right) dx$$

$$= \int \left(\sqrt{x} + \frac{1}{\sqrt{x}}\right) dx$$

$$= \int \left(x^{\frac{1}{2}} + x^{-\frac{1}{2}}\right) dx$$

$$= \frac{2}{3} x^{\frac{3}{2}} + 2x^{\frac{1}{2}} + C$$

The one on the left is not obvious – we can't use the same method because the function inside the square root is too complicated. So we change this by putting $u = x - 1$: then we have to think about the rest of the integral. The x-term on the top is OK because if $u = x - 1$, $x = u + 1$ and now our original integral:

$$\int \frac{x}{\sqrt{x-1}}\, dx \quad \text{has become} \quad \int \left(\frac{u+1}{\sqrt{u}}\right) dx$$

We can't integrate this as it stands, because the dx outside the bracket means that the integration is supposed to be with respect to the old variable, x, and we want to integrate with respect to u. Since:

$$dx = \frac{dx}{du} \times du$$

we can write $\frac{dx}{du}\, du$ instead of dx. We have to find $\frac{dx}{du}$ – easy in this case because $x = u + 1$ and $\frac{dx}{du} = 1$. So $dx = \frac{dx}{du} \times du = 1 \times du = du$

and our integral finally becomes:

$$\int \frac{u+1}{\sqrt{u}}\, du$$

which is the same as our other integral with u instead of x. We've already worked this out.

$$\int \frac{u+1}{\sqrt{u}}\, du = \frac{2}{3} u^{\frac{3}{2}} + 2u^{\frac{1}{2}} + C$$

Since our substitution was $x = u + 1$ we rearrange this to give $u = x - 1$ and put this back: we now have

$$\int \frac{x}{\sqrt{x-1}}\, dx = \frac{2}{3} u^{\frac{3}{2}} + 2u^{\frac{1}{2}} + C \quad \text{where } u = x - 1$$

$$= \frac{2}{3}(x-1)^{\frac{3}{2}} + 2(x-1)^{\frac{1}{2}} + C$$

Note that when we've finished transforming the original integral, everything must be in terms of the new variable – we can't end up with something like:

$$\int x^2 (1+u)\, dx \quad \text{or} \quad \int \frac{1+u^2}{x}\, du$$

We can, however, tolerate these mixtures of old and new while we're in the process of changing over: in fact, we shouldn't always be in too much of a hurry to substitute new for old – very frequently there can be cancellations or other simplifications. Let's have a look at an example of this.

Example

Find $\int x \sqrt{1 + 2x^2} \, dx$

Solution

Obviously we would like to simplify the function in the square root, so we put:

$$u = 1 + 2x^2$$

Now instead of worrying about putting the x to the left in terms of u, we'll go straight to dx. We know that $dx = \dfrac{dx}{du} \times du$, so we have to find $\dfrac{dx}{du}$.

To do this, we find $\dfrac{du}{dx}$ and use the fact that $\dfrac{dx}{du} = \dfrac{1}{\frac{du}{dx}}$

$$u = 1 + 2x^2 \quad \Rightarrow \quad \frac{du}{dx} = 4x$$

So: $\dfrac{dx}{du} = \dfrac{1}{4x}$,

and: $dx = \dfrac{dx}{du} \times du = \dfrac{1}{4x} \times du$

We'll put what we've worked out into our original integral and see what still needs doing.

$$\int x \sqrt{1 + 2x^2} \, dx = \int x \times \sqrt{u} \times \frac{1}{4x} \times du$$

We can see that there is a cancellation in this case, and we have finally

$$\int \frac{\sqrt{u}}{4} \, du = \int \frac{u^{\frac{1}{2}}}{4} \, du \quad = \frac{1}{6} u^{\frac{3}{2}} + C$$

$$= \frac{1}{6} (1 + 2x^2)^{\frac{3}{2}} + C$$

when we put back the original variable.

Example

Find $\int \dfrac{e^x}{(1 - 2e^x)^3} \, dx$

Solution

Put $u = 1 - 2e^x$

Then $\dfrac{du}{dx} = -2e^x$, $\dfrac{dx}{du} = -\dfrac{1}{2e^x}$ and $dx = \dfrac{dx}{du} \times du = -\dfrac{du}{2e^x}$

Our integral becomes: $\int \dfrac{e^x}{u^3} \times -\dfrac{du}{2e^x} = -\dfrac{1}{2} \int \dfrac{du}{u^3}$

$$= -\frac{1}{2} \int u^{-3} \, du$$

$$= \frac{1}{4} u^{-2} + C = \frac{1}{4u^2} + C$$

$$= \frac{1}{4(1 - 2e^x)^2} + C$$

Changing the limits

We need to know that if there are limits on our original integral, i.e. if it's a definite integral, then we have to change the limits in accordance with our substitution.

So for example with:

$$\int_0^1 x^4 (8 - 7x^5)^{\frac{2}{3}} \, dx$$

we would use the substitution $u = 8 - 7x^5$. The limits 1 and 0 refer to the x values. When $x = 1$, $u = 8 - 7 \times 1 = 1$ and when $x = 0$, $u = 8 - 7 \times 0 = 8$. The limits $x = 1$ and $x = 0$ are transformed to the limits $u = 1$ and $u = 8$ respectively – that is, the new limit occupies the same position as the corresponding old limit (even though this may mean the lower limit is larger than the upper, as in this case).

$$u = 8 - 7x^5 \text{ so } \frac{du}{dx} = -35x^4, \quad \frac{dx}{du} = -\frac{1}{35x^4}$$

$$dx = \frac{dx}{du} \times du = -\frac{1}{35x^4} \, du$$

$$\int_0^1 x^4 (8 - 7x^5)^{\frac{2}{3}} \, dx = \int_8^1 x^4 u^{\frac{2}{3}} \left(-\frac{1}{35x^4}\right) du$$

$$= -\frac{1}{35} \int_8^1 u^{\frac{2}{3}} \, du$$

$$= \frac{1}{35} \int_1^8 u^{\frac{2}{3}} \, du \quad \text{By property of limits}$$

$$= \frac{1}{35} \left[\frac{3}{5} u^{\frac{5}{3}} \right]_1^8$$

$$= \frac{1}{35} \times \frac{3}{5} \times \left(8^{\frac{5}{3}} - 1^{\frac{5}{3}} \right)$$

$$= \frac{3}{175} (32 - 1) = \frac{93}{175}$$

So when you use a substitution to evaluate a definite integral, remember you have to change three things:

1. the function you're integrating
2. the dx
3. the limits.

Note that, in the above example with a definite integral, we don't want to change back to the original variable.

Example

Find the exact value of

$$\int_{-\pi/2}^{\pi/2} \frac{\cos \theta}{\sqrt{5 + 3 \sin \theta}} \, d\theta$$

using the substitution $u = 5 + 3 \sin \theta$

Solution

When $\theta = \dfrac{\pi}{2}$, $u = 5 + 3\sin\dfrac{\pi}{2} = 8$

$\theta = -\dfrac{\pi}{2}$, $u = 5 + 3\sin\left(\dfrac{-\pi}{2}\right) = 2$

$\dfrac{du}{d\theta} = 3\cos\theta \Rightarrow d\theta = \dfrac{du}{3\cos\theta}$

Integral is now $\displaystyle\int_2^8 \dfrac{\cos\theta}{\sqrt{u}} \times \dfrac{du}{3\cos\theta}$

$= \dfrac{1}{3}\displaystyle\int_2^8 \dfrac{du}{\sqrt{u}} = \dfrac{1}{3}\left[2\sqrt{u}\right]_2^8$

$= \dfrac{2}{3}\left[\sqrt{8} - \sqrt{2}\right]$

$= \dfrac{2}{3}\sqrt{2}$

Practice questions E

1 Show that the following integrals have the stated values, using the suggested substitution

(a) $\displaystyle\int_0^{\pi/4} \dfrac{\sec^2\theta}{1+\tan\theta}\, d\theta = \ln 2 \; : \; u = 1 + \tan\theta$

(b) $\displaystyle\int_0^1 \dfrac{e^{2x}}{3e^{2x}+2}\, dx = \dfrac{1}{6}\ln\left(\dfrac{3e^2+2}{5}\right) \; :$

$u = 3e^{2x} + 2$

(c) $\displaystyle\int_0^{\sqrt{3}/2} \dfrac{x}{\sqrt{1-x^2}}\, dx = \dfrac{1}{2} \; : \; u = 1 - x^2$

(d) $\displaystyle\int_0^4 \dfrac{x+2}{\sqrt{2x+1}}\, dx = \dfrac{22}{3} \; : \; u = 2x + 1$

(e) $\displaystyle\int_0^2 \dfrac{x}{(x+1)^2}\, dx = \ln 3 - \dfrac{2}{3} \; : \; u = x + 1$

(f) $\displaystyle\int_0^{\pi/2} \sin x \sqrt{\cos x}\, dx = \dfrac{2}{3} \; : \; u = \cos x$

(g) $\displaystyle\int_2^4 \dfrac{\ln x}{x}\, dx = \dfrac{3}{2}(\ln 2)^2 \; : \; u = \ln x$

(h) $\displaystyle\int_0^1 \dfrac{x}{(3x+1)^2}\, dx = \dfrac{2}{9}\ln 2 - \dfrac{1}{12} \; :$

$n = 3x + 1$

2 (a) Use the substitution $u = e^x$ to show that

$\displaystyle\int_0^1 \dfrac{e^x - 1}{e^x + 1}\, dx = \int_1^e \dfrac{u - 1}{u(u+1)}\, du$

Hence, by using partial fractions, find the exact value of $\displaystyle\int_0^1 \dfrac{e^x - 1}{e^x + 1}\, dx$

3 The value of $\displaystyle\int_1^4 \dfrac{1}{x(1+\sqrt{x})}\, dx$ is denoted by I.

Use the substitution $u = \sqrt{x}$ to show that

$I = \displaystyle\int_1^2 \dfrac{2}{u(1+u)}\, du$

Hence, by using partial fractions, show that $I = \ln\left(\dfrac{16}{9}\right)$.

Integration by parts
OCR P3 5.3.5 (e)

We know the formula for differentiating a product, $uv \to uv' + vu'$, and we can apply it to $x \sin x$

$\dfrac{d(x \sin x)}{dx} = x \cos x + \sin x$

If $x \sin x$ differentiates to $x \cos x + \sin x$, it means that $x \cos x + \sin x$ integrates to $x \sin x$,

i.e. $\int (x \cos x + \sin x) \, dx = x \sin x + C$

Separating $\int x \cos x \, dx + \int \sin x \, dx = x \sin x + C$

$\Rightarrow \int x \cos x \, dx = x \sin x - \int \sin x \, dx + C$

$= x \sin x + \cos x + C$

By this indirect means we have managed to integrate a product, $x \cos x$, which we would otherwise be unable to do. In general:

$$\frac{d(uv)}{dx} = \frac{u \, dv}{dx} + \frac{v \, du}{dx}$$

Then if uv differentiates to the RHS, the RHS integrates to uv,

i.e. $\int u \frac{dv}{dx} \, dx + \int v \frac{du}{dx} \, dx = uv$

Rearranging this,

$$\int u \frac{dv}{dx} \, dx = uv - \int v \frac{du}{dx} \, dx$$

This means that if we have a product to integrate, we can change the integral into another using this formula, called the 'by parts' formula. We differentiate one term, $u \to \frac{du}{dx}$ and integrate the other $\frac{dv}{dx} \to v$

The new product, if we have chosen u and $\frac{dv}{dx}$ wisely, can turn out to be easier to integrate. Let's have a look at an example of this:

Example Evaluate $\int x \cos 2x \, dx$.

Solution Put $u = x$ and $\frac{dv}{dx} = \cos 2x$

$\therefore \frac{du}{dx} = 1$ and $v = \frac{1}{2} \sin 2x$ (ignore $+ C$)

Now substitute in the 'by parts' formula to get:

$\int x \cos 2x \, dx = \frac{x}{2} \sin 2x - \int \frac{1}{2} \sin 2x \, dx$

$= \frac{x}{2} \sin 2x - \frac{1}{2} \int \sin 2x \, dx$ (Take constants outside the integral if possible)

$\therefore \int x \cos 2x \, dx = \frac{x}{2} \sin 2x + \frac{1}{4} \cos 2x + C$

Example

Find $\int xe^{-2x}\,dx$.

Solution

Put $u = x$ and $\dfrac{dv}{dx} = e^{-2x}$

$\therefore\quad \dfrac{du}{dx} = 1$ and $v = -\dfrac{1}{2}e^{-2x}$

Now substitute in the 'by parts' formula to get:

$$\int xe^{-2x}\,dx = -\dfrac{x}{2}e^{-2x} - \int -\dfrac{1}{2}e^{-2x}\,dx$$

$$= -\dfrac{x}{2}e^{-2x} + \dfrac{1}{2}\int e^{-2x}\,dx. \text{ (Always simplify)}$$

$\therefore\quad \int xe^{-2x}\,dx = -\dfrac{x}{2}e^{-2x} - \dfrac{1}{4}e^{-2x} + C$

Note that in both these examples we have taken u to be x; this is mostly true, except in the case where one of the terms of the produce is $\ln x$. In this case, we take u to be $\ln x$ and integrate the power of x.

Example

Find $\int x^5 \ln x\,dx$

Solution

$u = \ln x$ and $\dfrac{dv}{dx} = x^5$

then $\dfrac{du}{dx} = \dfrac{1}{x}$ and $v = \dfrac{x^6}{6}$

i.e. $\int x^5 \ln x\,dx = \dfrac{x^6}{6}\ln x - \int \dfrac{1}{x} \times \dfrac{x^6}{6}\,dx$

$$= \dfrac{x^6}{6}\ln x - \dfrac{1}{6}\int x^5\,dx$$

$$= \dfrac{x^6}{6}\ln x - \dfrac{1}{36}x^6 + C$$

$$= \dfrac{x^6}{36}(6\ln x - 1) + C$$

Integration by parts – twice

OCR P3 5.3.5 (e)

When the x term turns out to be x^2, we can sometimes find the integral by integrating twice by parts, for example:

$\int x^2 e^{2x}\,dx \qquad\qquad u = x^2 \qquad \dfrac{dv}{dx} = e^{2x}$

$= \dfrac{1}{2}x^2 e^{2x} - \int 2x \times \dfrac{1}{2}e^{2x}\,dx \qquad \dfrac{du}{dx} = 2x \qquad v = \dfrac{1}{2}e^{2x}$

$= \dfrac{1}{2}x^2 e^{2x} - \int xe^{2x}\,dx \quad\leftarrow$ We now find this integral using the by parts method

$u = x \qquad \dfrac{dv}{dx} = e^{2x}$

$\dfrac{du}{dx} = 1 \qquad v = \dfrac{1}{2}e^{2x}$

and our original integral becomes

$$\frac{1}{2}x^2 e^{2x} - \left(\frac{1}{2}xe^{2x} - \frac{1}{2}\int e^{2x}\right)$$

$$= \frac{1}{2}x^2 e^{2x} - \frac{1}{2}xe^{2x} + \frac{1}{4}e^{2x} + C$$

$$= \frac{e^{2x}}{4}(2x^2 - 2x + 1) + C$$

There is one other 'trick' that you have to know. Sometimes a single function occurs and you still integrate as a product; the second term of the product is taken as 1:

e.g. $\int \ln x \, dx$ $u = \ln x$ $\frac{dv}{dx} = 1$

$= \int (\ln x \times 1) \, dx$ $\frac{du}{dx} = \frac{1}{x}$ $v = x$

$= x \ln x - \int \frac{1}{x} \times x \, dx$

$= x \ln x - x + C$

Practice questions F

1 Integrate the following using parts:

(a) $\int xe^{-x} \, dx$ (b) $\int xe^{3x} \, dx$

(c) $\int x \cos \frac{x}{2} \, dx$ (d) $\int x^2 \ln x \, dx$

(e) $\int x \sin x \, dx$ (f) $\int x \ln 2x \, dx$

(g) $\int x(2x-1)^5 \, dx$ (h) $\int \ln 3x \, dx$

2 Show the following using the by parts method:

(a) $\int_0^1 \frac{x}{e^x} \, dx = \frac{e-2}{e}$

(b) $\int_1^e x^3 \ln x \, dx = \frac{1}{16}(3e^4 - 1)$

(c) $\int_0^{\pi/2} x \cos 2x \, dx = \frac{-1}{2}$

(d) $\int_0^{\pi/3} x \sin 3x \, dx = \frac{\pi}{9}$

3 Find the following, using parts twice:

(a) $\int x^2 \sin 2x \, dx$ (b) $\int x^2 e^{2x} \, dx$

(c) $\int_0^{\pi/3} x^2 \cos 3x \, dx$ (d) $\int_0^{\pi} x^2 \sin x \, dx$

4 Show that:

$$\int_0^1 (x+1)e^{-x} \, dx = 2 - \frac{3}{e}$$

5 Find $\int xe^{2x} \, dx$. Hence show that:

$$\int x^2 e^{2x} \, dx = \left(\frac{1}{2}x^2 - \frac{1}{2}x + \frac{1}{4}\right) e^{2x} + C,$$

where C is an arbitrary constant.

6 Integrate by parts twice (differentiating the function involving $\ln x$ in each case) to show that:

$$\int x^2 (\ln x)^2 \, dx = \frac{x^3}{3}(\ln x)^2 - \frac{2}{9}x^3 \ln x + \frac{2}{27}x^3 + C$$

7 Show that:

$$\int (\ln x)^2 \, dx = x(\ln x)^2 - 2x \ln x + 2x + C$$

8 (a) Show that $\frac{x^3}{x^2 + 1} = x - \frac{x}{x^2 + 1}$

(b) Hence show that:

$$\int_1^2 x \ln(x^2 + 1) \, dx = \frac{5}{2} \ln 5 - \ln 2 - \frac{3}{2}$$

9 Show that $\int_0^{\pi/2} x^2 \sin x \cos x \, dx = \frac{1}{16}\pi^2 - \frac{1}{4}$

10 Show that:

$$\int x \tan^2 x \, dx = x \tan x + \ln \cos x - \frac{1}{2}x^2 + C$$

P3 Section 5

11 Find the following:

(a) $\int_0^{2/3} (3x-2)^3 \, dx$ (b) $\int x \sin 2x \, dx$

(c) $\int \dfrac{1}{x(x+1)} \, dx$ (d) $\int (e^x - 1)^2 \, dx$

(e) $\int_1^2 (x+1)\sqrt{2-x} \, dx$

(f) $\int_1^e x^2 \ln x \, dx$ (g) $\int \dfrac{x}{4+x} \, dx$

(h) $\int (2x+3) \ln x \, dx$ (i) $\int \dfrac{\cos x}{3 + \sin x} \, dx$

(j) $\int_0^{\pi/4} \sin 3x \sin 2x \, dx$ (k) $\int \dfrac{x^3}{x^2+1} \, dx$

(l) $\int_0^{\pi/2} x^2 \cos x \, dx$ (m) $\int \dfrac{x^2 + 5}{(x-1)(x-2)} \, dx$

(n) $\int_0^{\pi/2} \cos^4 x \sin x \, dx$ (o) $\int_0^3 \dfrac{2x+1}{\sqrt{x+1}} \, dx$

(p) $\int \cos\theta \sqrt{1 + \sin\theta} \, d\theta$

(q) $\int x \ln\left(\dfrac{1}{x}\right) dx$ (r) $\int_1^2 \dfrac{6}{9-x^2} \, dx$

(s) $\int \dfrac{x+2}{(x^2+1)(2x-1)} \, dx$

(t) $\int_1^e \dfrac{\ln x}{\sqrt{x}} \, dx$

12 Show that $\dfrac{d}{dx}\left(\dfrac{x^3}{1+x^4}\right) = \dfrac{x^2(3-x^4)}{(1+x^4)^2}$

Hence find $\int_1^2 \dfrac{x^2(3-x^4)}{(1+x^4)^2} \, dx$

Trigonometric integration

OCR P3 5.3.5 (a)

There is a certain type of expression that we can integrate directly by recognising that it comes from differentiating a power of a trigonometric function.

Suppose we differentiate $\sin^5 x$

$\sin^5 x = (\sin x)^5 \rightarrow 5(\sin x)^4 \cos x$
$\qquad = 5 \sin^4 x \cos x$

Reversing this means that $\int 5 \sin^4 x \cos x \, dx = \sin^5 x + C$

or $\int \sin^4 x \cos x \, dx = \dfrac{1}{5} \sin^5 x + D$ on dividing by 5.

Similarly, $\tan^3 x = (\tan x)^3$ differentiates to $3 \tan^2 x \sec^2 x$, and so

$\int 3 \tan^2 x \sec^2 x \, dx = \tan^3 x + C$

or $\int \tan^2 x \sec^2 x \, dx = \dfrac{1}{3} \tan^3 x + D$

We can see the pattern: when we have a power of a function multiplied by the derivative, something like $\int \square^n \blacksquare$, we can integrate by treating the function as though it were just x,

i.e. $\int \square^n \blacksquare = \dfrac{\square^{n+1}}{n+1} + C$

so that $\int \sin^2 x \cos x \, dx = \dfrac{1}{3} \sin^3 x + C$

$\int \tan^7 x \sec^2 x \, dx = \dfrac{1}{8} \tan^8 x + C$

$\int \cos^4 x \sin x \, dx = -\dfrac{1}{5} \cos^5 x + C$

(since $\sin x$ is not exactly the derivative of $\cos x$: we need $-\sin x$)

Integration P3

Practice questions G

1 Write down the following

(a) $\int \sin^3 x \cos x \, dx$
(b) $\int \tan^4 x \sec^2 x \, dx$
(c) $\int \cos x \sin x \, dx$
(d) $\int \frac{\cos x}{\sin^2 x} \, dx$
(e) $\int \sqrt{\tan x} \sec^2 x \, dx$
(f) $\int \frac{\sin x}{\sqrt{\cos x}} \, dx$

Using identities

OCR P3 5.3.5 (b)

Sometimes we may need to apply the formulae we looked at in unit P2. For reference, some of these are:

$$\sin^2 \theta + \cos^2 \theta = 1$$
$$\cos 2\theta = \cos^2 \theta - \sin^2 \theta$$
$$\sin 2\theta = 2 \sin \theta \cos \theta$$

From the first equation we can find either $\sin^2 \theta$ or $\cos^2 \theta$ in terms of the other and substitute this into the second equation to give:

$$\cos^2 \theta = 1 - \sin^2 \theta \qquad\qquad \sin^2 \theta = 1 - \cos^2 \theta$$
$$\Rightarrow \cos 2\theta = (1 - \sin^2 \theta) - \sin^2 \theta \Rightarrow \cos 2\theta = \cos^2 \theta - (1 - \cos^2 \theta)$$
$$= 1 - 2\sin^2 \theta \qquad\qquad = 2\cos^2 \theta - 1$$

These are useful formulae in their own right, but for integration purposes we rewrite them:

$$\sin^2 \theta = \tfrac{1}{2}(1 - \cos 2\theta) \;;\; \cos^2 \theta = \tfrac{1}{2}(1 + \cos 2\theta)$$

θ can be any angle we like, so, for example:

$$\sin^2 5x = \tfrac{1}{2}(1 - \cos 10x) \text{ and } \cos^2 \tfrac{y}{8} = \tfrac{1}{2}\left(1 + \cos \tfrac{y}{4}\right)$$

The advantage lies in transforming something involving a power, which can't be integrated immediately, into something involving a multiple angle, which can be integrated directly.

Example

Evaluate: $\int_0^{\frac{\pi}{4}} \cos^2 4x \, dx$

Solution

Using our formula, we find that:

$$\cos^2 4x = \tfrac{1}{2}(1 + \cos 8x)$$

Our integral becomes:

$$\int_0^{\frac{\pi}{4}} \cos^2 4x \, dx = \int_0^{\frac{\pi}{4}} \tfrac{1}{2}(1 + \cos 8x) \, dx$$

$$= \left[\tfrac{1}{2}\left(x + \frac{\sin 8x}{8}\right)\right]_0^{\frac{\pi}{4}}$$

$$= \tfrac{1}{2}\left(\frac{\pi}{4} + \frac{\sin 2\pi}{8}\right) = \frac{\pi}{8} \text{ (since } \sin 2\pi = 0\text{)}$$

Things can be easier with the other ratios because some of their derivatives already involve squares:

since $\dfrac{d(\tan x)}{dx} = \sec^2 x$, for example, we have

$$\int \sec^2 x \, dx = \tan x + C$$

we can then use the Pythagorean identity

$$1 + \tan^2 x = \sec^2 x$$

to integrate $\tan^2 x$

$$\int \tan^2 x = \int (\sec^2 x - 1) \, dx$$
$$= \tan x - x + C$$

Similarly with the remaining two:

$$\dfrac{d(\cot x)}{dx} = -\operatorname{cosec}^2 x, \text{ so}$$

$$\int \operatorname{cosec}^2 x \, dx = -\cot x + C$$

and $\int \cot^2 x \, dx = \int (\operatorname{cosec}^2 x - 1) \, dx$
$$= -\cot x - x + C$$

We can use this to find higher powers

$$\begin{aligned}\int \tan^4 x \, dx &= \int \tan^2 x \cdot \tan^2 x \, dx \\ &= \int \tan^2 x \cdot (\sec^2 x - 1) \, dx \\ &= \int \tan^2 x \cdot \sec^2 x - \int \tan^2 x \, dx \\ &= \int \tan^2 x \sec^2 x - \int (\sec^2 x - 1) \, dx \\ &= \int \tfrac{1}{3} \tan^3 x - \tan x + x + C \end{aligned}$$

Integration of trigonometric functions is a massive subject and this is only a very brief introduction. There is a final type that you need to be able to integrate which comes from the factor formulae we looked at in P2: these are:

$$\begin{aligned}\sin(A+B) + \sin(A-B) &\equiv 2 \sin A \cos B \\ \sin(A+B) - \sin(A-B) &\equiv 2 \cos A \sin B \\ \cos(A+B) + \cos(A-B) &\equiv 2 \cos A \cos B \\ \cos(A+B) - \cos(A-B) &\equiv -2 \sin A \sin B \end{aligned}$$

So to tackle an integral like $\int \cos 3x \cos 5x \, dx$, we see that this corresponds to the third identity with $A = 3x$ and $B = 5x$.

$$\begin{aligned}\int \cos 3x \cos 5x \, dx &= \tfrac{1}{2} \int \cos 8x + \cos(-2x) \, dx \\ &= \tfrac{1}{2} \left[\tfrac{1}{8} \sin 8x + \tfrac{1}{2} \sin 2x \right] + C\end{aligned}$$

(Since cos is an even function, $\cos(-2x) = \cos 2x$.)

Example

Find $\displaystyle\int_0^{\pi/2} \sin 4x \cos x \, dx$

Solution

This corresponds to the first identity with $A = 4x$, $B = x$

$$\int_0^{\pi/2} \sin 4x \cos x \, dx = \frac{1}{2} \int_0^{\pi/2} \sin 5x + \sin 3x \, . \, dx$$

$$= \frac{1}{2}\left[-\frac{1}{5} \cos 5x - \frac{1}{3} \cos 3x \right]_0^{\pi/2}$$

$$= \frac{1}{2}\left[0 - \left(-\frac{1}{5} - \frac{1}{3} \right) \right]$$

$$= \frac{4}{15}$$

Practice questions H

1. Find the following
   (a) $\int \cos^2 3x \, dx$
   (b) $\int \sin^2 \left(\frac{x}{3}\right) dx$
   (c) $\int 3 \tan^2 x \, dx$
   (d) $\int \sec^4 x \, dx$

2. Show that $\int_0^{\pi/4} (1 - \tan x)^2 \, dx = 1 - \ln 2$

3. Find $\int \cos 5x \sin 2x \, dx$

4. Evaluate $\int_0^{\pi/4} \sin 5x \cos 3x \, dx$

5. Differentiate $\tan^2 x$, writing your answer in the form $a(\tan x + \tan^3 x)$

   Hence show that $\int_0^{\pi/4} (\tan x + \tan^3 x) \, dx = \frac{1}{2}$

   and use this to evaluate $\int_0^{\pi/4} \tan^3 x \, dx$

Areas

We are going to look at the area under a curve with equation in parametric form. Essentially, we use the same formulae as we would for equations in cartesian form, i.e.

$$\text{x-axis, } \int y \, dx \text{ or y-axis, } \int x \, dy$$

and proceed as if we were making a substitution, changing all the variables, including the limits, to the parameter.

i.e. $\int_{x_1}^{x_2} y \, dx \equiv \int_{t_1}^{t_2} y \frac{dy}{dt} \, dt \quad$ with y in terms of t

or $\int_{y_1}^{y_2} x \, dy \equiv \int_{t_1}^{t_2} x \frac{dy}{dt} \, dt \quad$ with x in terms of t

Suppose we had to find the area under the curve given by $x = t^2 - 1$, $y = 2t + 1$ which lies between the lines $x = 3$ and $x = 0$ and above the x-axis. Fig. 5.1 is a sketch of this:

Figure 5.1

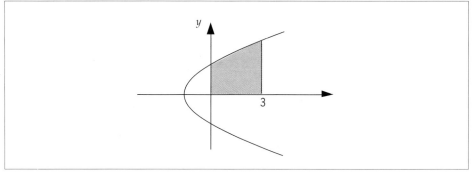

The limits $x = 3$ and $x = 0$, using $t = \sqrt{x+1}$, become $t = 2$ and $t = 1$ (taking the positive values since the curve is above the x-axis).

y is simply replaced by $2t + 1$.

Since $x = t^2 - 1$, $\dfrac{dx}{dt} = 2t$ and the integral becomes:

$$\int_1^2 (2t+1) \, 2t \, dt = \int_1^2 (4t^2 + 2t) \, dt = \left[\frac{4t^3}{3} + t^2\right]_1^2$$

$$= \left[\left(\frac{32}{3} + 4\right) - \left(\frac{4}{3} + 1\right)\right] = \frac{37}{3}$$

We could also have found the area by eliminating t for the corresponding cartesian equation and integrating as normal.

Here's an example where the parametric equations involve trigonometric functions.

Example

A curve is defined by the parametric equations $x = a \cos t$, $y = b \sin t$, $0 \leq t \leq 2\pi$. Find the area in the first quadrant enclosed by the curve, the y-axis and the line $y = \dfrac{b}{2}$, shaded in the sketch.

Figure 5.2

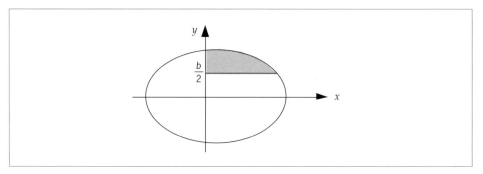

Solution

We are going to use the area between the curve and the y-axis,

i.e. $\displaystyle\int_{y_1}^{y_2} x \, dy$. Limits are $y = b$ and $y = \dfrac{b}{2}$

but substitute: when $y = b$, $b \sin t = b \Rightarrow t = \dfrac{\pi}{2}$

$y = \dfrac{b}{2}$, $b \sin t = \dfrac{b}{2} \Rightarrow t = \dfrac{\pi}{6}$

$$x = a\cos t, \quad \frac{dy}{dt} = b\cos t$$

$$\Rightarrow \text{integral becomes} \int_{\pi/6}^{\pi/2} (a\cos t)(b\cos t)\,dt$$

$$= ab \int_{\pi/6}^{\pi/2} \cos^2 t\,dt = \frac{ab}{2} \int_{\pi/6}^{\pi/2} (1 + \cos 2t)\,dt$$

$$= \frac{ab}{2}\left[t + \frac{1}{2}\sin 2t\right]_{\pi/6}^{\pi/2} = \frac{ab}{2}\left[\frac{\pi}{2} - \frac{\pi}{6} - \frac{\sqrt{3}}{4}\right]$$

$$= \frac{ab}{2}\left[\frac{\pi}{3} - \frac{\sqrt{3}}{4}\right] = \frac{ab}{24}(4\pi - 3\sqrt{3})$$

Practice questions I

1 Find the following areas, using the parametric equations

(a)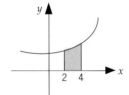

$x = 2t,\ y = t^2 + 1$

between $x = 2$, $x = 4$

(b)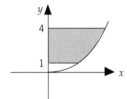

$x = t^3,\ y = t^2$

between $y = 1$, $y = 4$

(c)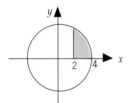

$x = 4\cos\theta,$
$y = 4\sin\theta$

between $x = 4$ and $x = 2$

(d)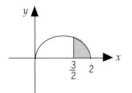

$x = 1 + \cos\theta,$
$y = 2\sin^2\theta$

for $0 \le \theta \le \pi$

between $x = \frac{3}{2}$ and $x = 2$

2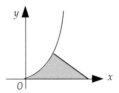

The curve shown in the above figure has parametric equations

$$x = t^2,\ y = t^3$$

where $t \ge 0$ is a parameter. Also shown is part of the normal to the curve at the point where $t = 1$.

(a) Find an equation of this normal

(b) Find the area of the finite region bounded by the curve, the x-axis and this normal.

3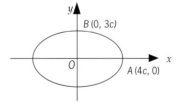

The above figure shows a sketch of the ellipse given parametrically by the equations

$$x = 4c\cos t,\ y = 3c\sin t,\ -\pi < t \le \pi$$

where c is a positive constant.

(a) Write down the value of t at the point $A(4c, 0)$ and at the point $B(0, 3c)$

(b) By considering the integral $\int y\,\frac{dx}{dt}\,dt$ find, in terms of c, the area of the region enclosed by the ellipse.

4 Sketch the curve given parametrically by
$$x = t^2, \quad y = t^3$$
Show that an equation of the normal to the curve at the point A (4, 8) is $x + 3y - 28 = 0$

This normal meets the x-axis at the point N. Find the area of the region enclosed by the arc OA of the curve, the line segment AN and the x-axis.

5

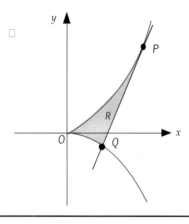

The curve C has two arcs, as shown in the above figure, and the equations
$$x = 3t^2, \quad y = 2t^3$$
where t is a parameter.

(a) Find an equation of the tangent to C at the point P where $t = 2$.

The tangent meets the curve again at the point Q.

(b) Show that the coordinates of Q are $(3, -2)$.

The shaded region R is bounded by the arcs OP and OQ of the curve C, and the line PQ, as shown in the figure.

(c) Find the area of R.

SUMMARY EXERCISE

1 (a) Find the value of $\displaystyle\int_0^{\frac{1}{4}\pi} \sec^2 x \, dx$

(b) A region R in the first quadrant is bounded by the curve $y = \tan x$, the x-axis and the line $x = \frac{1}{4}\pi$. Show that the exact value of the volume of the solid formed when R is rotated completely about the x-axis is $\pi - \frac{1}{4}\pi^2$.

2

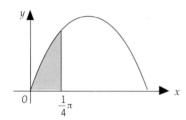

The region R, shown shaded in the diagram, is bounded by the curve $y = \sin x$, the x-axis and the line $x = \frac{1}{4}\pi$. Find the exact value of the volume of the solid formed when R is rotated completely about the x-axis.

3 Using integration by parts, show that
$$\int_0^{\pi/4} x \sec^2 x \, dx = \frac{1}{4}(\pi - \ln 4)$$

4 Use the substitution $u = \ln x$ to show that
$$\int_e^{e^2} \frac{1}{\sqrt{(\ln x)}} \, dx = 2\sqrt{2} - 2$$

5 By using the substitution $u = \sin x$, or otherwise, find
$$\int \sin^3 x \sin 2x \, dx$$
giving your answer in terms of x.

6 By means of the substitution $u = 3 + e^{-x}$, find the exact value of
$$\int_{\ln 0.5}^{0} \frac{e^{-x}}{2\sqrt{(3 + e^{-x})}} \, dx$$

7 Use the substitution $u = 4 + x^2$ to show that
$$\int_0^1 \frac{x^3}{4 + x^2} \, dx = \frac{1}{3}(16 - 7\sqrt{5})$$

8 Show that $\cos 6x + \cos 4x = 2 \cos 5x \cos x$

Find $\displaystyle\int_0^{\pi/12} \cos 5x \cos x \, dx$, giving your solution as an exact value.

9 (a) Differentiate $\ln(x^3 + 6x)$ with respect to x

(b) Show that $\displaystyle\int_2^3 \frac{x^2+2}{x^3+6x} \, dx = \frac{2}{3} \ln \frac{3}{2}$

10 Show that $\displaystyle\int_0^{\pi/2} x^2 \sin 2x \, dx = \frac{1}{8}\pi^2 - \frac{1}{2}$

The region between the curve with equation $y = x(\cos x + \sin x)$, the x-axis and the lines $x = 0$ and $x = \frac{1}{2}\pi$ is rotated completely about the x-axis. Show that the volume V of the solid so formed is given by

$$V = \pi \int_0^{\frac{1}{2}\pi} x^2(1 + \sin 2x) \, dx$$

and evaluate V, giving your answer in terms of π.

11 (a) Show that $\displaystyle\int \sin^2 x \, dx = \frac{1}{2}x - \frac{1}{4}\sin 2x - c$ where c is an arbitrary constant.

(b) Show that $\displaystyle\int_0^{\pi/2} x \sin^2 x \, dx = \frac{1}{16}(\pi^2 + 4)$

12 Express $\dfrac{6x+4}{(1-2x)(1+3x^2)}$ in partial fractions.

Hence show that

$$\int_1^2 \frac{6x+4}{(1-2x)(1+3x^2)} \, dx = \ln\left(\frac{13}{36}\right)$$

13 $f(x) \equiv \dfrac{5x^2 - 8x + 1}{2x(x-1)^2} \equiv \dfrac{A}{x} + \dfrac{B}{x-1} + \dfrac{C}{(x-1)^2}$

(a) Find the values of the constants A, B and C

(b) Hence find $\int f(x) \, dx$

(c) Hence show that

$$\int_4^9 f(x) \, dx = \ln\left(\frac{32}{3}\right) - \frac{5}{24}$$

14 (a) Find $\int x(x^2 + 3)^5 \, dx$

(b) Show that $\displaystyle\int_1^e \frac{1}{x^2} \ln x \, dx = 1 - \frac{2}{e}$

(c) Given that $p > 1$, show that

$$\int_1^p \frac{1}{(x+1)(2x-1)} \, dx = \frac{1}{3} \ln \frac{4p-2}{p+1}$$

15

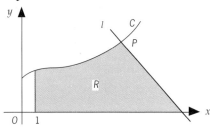

The above figure shows part of the curve C with parametric equations.

$$x = (t+1)^2, \quad y = \frac{1}{2}t^3 + 3, \quad t \geq -1$$

P is the point on the curve where $t = 2$.
The line l is the normal to C at P.

(a) Find an equation of l

The shaded region R is bounded by C, l, the x-axis and the line with equation $x = 1$.

(b) Using integration and showing all your working, find the area of R.

16

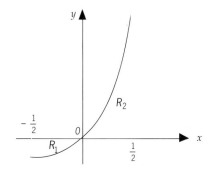

The above figure shows the curve with equation $y = xe^{2x}$, $-\frac{1}{2} \leq x \leq \frac{1}{2}$.

The finite region R_1 bounded by the curve, the x-axis and the line $x = -\frac{1}{2}$, has area A_1.

The finite region R_2 bounded by the curve, the x-axis and the line $x = \frac{1}{2}$, has area A_2.

(a) Find the exact values of A_1 and A_2 by integration

(b) Show that $A_1 : A_2 = (e-2) : e$.

17

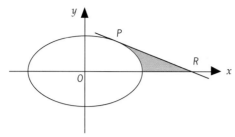

The curve in the above figure has parametric equations

$$x = 5\cos\theta, \quad y = 4\sin\theta, \quad 0 \le \theta < 2\pi$$

(a) Find the gradient of the curve at the point P at which $\theta = \dfrac{\pi}{4}$.

(b) Find an equation of the tangent to the curve at the point P.

(c) Find the coordinates of the point R where this tangent meets the x-axis.

The shaded region, shown in the figure is bounded by the tangent PR, the curve and the x-axis.

(d) Find the area of the shaded region, leaving your answer in terms of π.

SUMMARY

When you have finished this section, you should:

- know the integral of $\sin x$, $\cos x$ and $\sec^2 x$
- know the integral of related functions, e.g. $\sin 2x$
- recognise an integral of the type derivative ÷ function and be able to carry out the integration, adjusting the constants
- recognise and be able to integrate functions of the form $\dfrac{1}{ax+b}$ or $(ax+b)^n$
- be aware that you can use partial fractions to integrate certain rational functions
- be particularly familiar with the three types of functions arising from partial fractions and their integrals
- know that certain integrals can be simplified by changing the variable with a substitution
- be familiar with the process of changing the variable and what it entails
- be able to choose the substitution in simple cases
- be familiar with the formula for integration by parts and its derivation from the product rule of differentiation
- be able to carry out the process for standard products of the form $x\,f(x)$
- know that you have to repeat the process if the function is of the form $x^2 f(x)$
- know that a function of the form $x^n \ln x$ is integrated by parts by differentiating the $\ln x$ (including the case where $n = 0$)
- know how to integrate trigonometric functions of the form $[f(x)]^n f'(x)$
- know how to use trigonometric identities to simplify the integration of functions such as $\cos^2 x$, $\tan^2 x$
- know that products of sine and/or cosine functions can sometimes be simplified using the factor formulae from unit P2
- be able to integrate functions of either of the two types above
- know how to integrate functions given in parametric form to find the area between a curve and the x- or y-axis.

ANSWERS

Practice questions A

1. (a) $\dfrac{-1}{2}\cos 2x + C$ (b) $\dfrac{1}{3}\sin 3x + C$

   (c) $\dfrac{1}{5}\tan 5x + C$ (d) $\dfrac{1}{3}e^x + C$

   (e) $\dfrac{1}{2}\sin 4x + C$ (f) $\cos 3x + C$

   (g) $6\tan\dfrac{x}{3} + C$ (h) $\dfrac{-1}{4}e^{-4x} + C$

   (i) $\cos\left(\dfrac{\pi}{6} - x\right) + C$ (j) $\dfrac{2}{3}\left(3x - \dfrac{\pi}{4}\right) + C$

   (k) $-2e^{3-\frac{1}{2}x} + C$ (l) $-8\cos\dfrac{x}{2} + C$

Practice questions B

1. (a) $\ln|1+x| + C$
   (b) $\dfrac{1}{3}\ln|1+3x| + C$
   (c) $2\ln|x+4| + C$
   (d) $\dfrac{3}{2}\ln|1+2x| + C$
   (e) $\dfrac{-1}{2}\ln|1-4x| + C$
   (f) $\dfrac{-1}{2}\ln|3-10x| + C$
   (g) $\dfrac{7}{3}\ln|3x+2| + C$
   (h) $\dfrac{2}{3}\ln|2+9x| + C$

2. (a) $\dfrac{1}{4}\ln(1+x^4) + C$
   (b) $\dfrac{-1}{6}\ln|1-2x^3| + C$
   (c) $\dfrac{-1}{2}\ln|\cos 2x| + C$
   (d) $\dfrac{1}{3}\ln|\sin 3x| + C$
   (e) $\dfrac{-1}{2}\ln|1-x^2| + C$
   (f) $\dfrac{3}{4}\ln(1+2x^2) + C$
   (g) $-2\ln|1-e^x| + C$
   (h) $\ln(x^2-3x+9) + C$
   (i) $\ln|1+\tan x| + C$
   (j) $\ln(2x-5)(x+3) + C$
   (k) $\ln(\ln x) + C$
   (l) $2\ln(1+\sqrt{x}) + C$

Practice questions C

1. (a) $\dfrac{(3x+2)^5}{15} + C$ (b) $-\dfrac{1}{1+x} + C$

   (c) $\dfrac{1}{6}(1+4x)^{\frac{3}{2}} + C$ (d) $\dfrac{2}{3(1-3x)^2}$

   (e) $\dfrac{1}{2(3-x)} + C$ (f) $-3(3-2x)^{\frac{2}{3}} + C$

Practice questions D

1. (a) $\ln\left|\dfrac{x-3}{x+1}\right| + C$

   (b) $\ln\left|\dfrac{(x+1)(x-1)}{x^2}\right| + C$

   (c) $x + \ln\left|\dfrac{x+2}{x+1}\right| + C$

   (d) $\ln\left|\dfrac{x-1}{2x+1}\right| + \dfrac{1}{x-1} + C$

   (e) $\ln\left|\dfrac{x-3}{2x^2+1}\right| + C$

   (f) $\ln\left|\dfrac{x-2}{x+2}\right| + C$

2. (a) $\ln\dfrac{5}{2} + \dfrac{1}{2}$ (b) $\ln\dfrac{4}{3} - \dfrac{1}{6}$

   (c) $\ln 5$ (d) $\dfrac{1}{2}\ln 7 - \ln 3$

Practice questions E

2. $2\ln\left(\dfrac{e+1}{2}\right) - 1$

Practice questions F

1. (a) $-(x+1)e^{-x} + C$ (b) $\dfrac{1}{3}xe^{3x} - \dfrac{1}{9}e^{3x} + C$

   (c) $2x\sin\dfrac{1}{2}x + 4\cos\dfrac{1}{2}x + C$

   (d) $\dfrac{1}{3}x^3\ln x - \dfrac{1}{9}x^3 + C$

   (e) $\sin x - x\cos x + C$

   (f) $\dfrac{1}{4}x^2(2\ln 2x - 1) + C$

   (g) $\dfrac{1}{168}(2x-1)^6(12x+1) + C$

   (h) $x\ln 3x - x + C$

Section 5

3 (a) $\dfrac{-x^2 \cos 2x}{2} + \dfrac{x \sin 2x}{2} + \dfrac{\cos 2x}{4} + C$

  (b) $\dfrac{1}{4} e^{2x}(2x^2 - 2x + 1) + C$

  (c) $\dfrac{-2\pi}{27}$ (d) $\pi^2 - 4$

5 $\dfrac{1}{2} x e^{2x} - \dfrac{1}{4} e^{2x} + C$

11 (a) $\dfrac{-4}{3}$ (b) $\dfrac{-1}{2} x \cos 2x + \dfrac{1}{4} \sin 2x + C$

  (c) $\ln \left| \dfrac{x}{x+1} \right| + C$

  (d) $\dfrac{1}{2} e^{2x} - 2e^x + x + C$ (e) $\dfrac{8}{5}$

  (f) $\dfrac{2}{3} e^3 + \dfrac{1}{9}$ (g) $x - 4\ln(4+x) + C$

  (h) $(x^2 + 3x)\ln x - \dfrac{1}{2} x^2 - 3x + C$

  (i) $\ln(3 + \sin x) + C$ (j) $\dfrac{3\sqrt{2}}{10}$

  (k) $\dfrac{x^2}{2} - \dfrac{1}{2} \ln(x^2 + 1) + C$ (l) $\dfrac{1}{4} \pi^2 - 2$

  (m) $x + 9\ln(x-2) - 6\ln(x-1)$

  (n) $\dfrac{1}{5}$ (o) $\dfrac{22}{3}$ (p) $\dfrac{2}{3}(1 + \sin \theta)^{\frac{3}{2}} + C$

  (q) $\dfrac{-1}{2} x^2 \ln x + \dfrac{1}{4} x^2 + C$ (r) $\ln \dfrac{5}{2}$

  (s) $\ln \dfrac{2x-1}{\sqrt{x^2+1}} + C$ (t) $4 - 2e^{\frac{1}{2}}$

12 $\dfrac{-1}{34}$

Practice questions G

1 (a) $\dfrac{\sin^4 x}{4} + C$ (b) $\dfrac{\tan^5 x}{5} + C$

  (c) $-\dfrac{\cos^2 x}{2} + C$ or $\dfrac{\sin^2 x}{2} + C$

  (d) $-\dfrac{1}{\sin x} + C$ (e) $\dfrac{2}{3}(\tan x)^{\frac{3}{2}}$

  (f) $-2\sqrt{\cos x} + C$

Practice questions H

1 (a) $\dfrac{1}{2} x + \dfrac{1}{12} \sin 6x + C$

  (b) $\dfrac{1}{2} x - \dfrac{3}{4} \sin\left(\dfrac{2x}{3}\right) + C$

  (c) $3(\tan x - x) + C$

  (d) $\tan x + \dfrac{1}{3} \tan^3 x + C$

3 $\dfrac{-1}{14} \cos 7x + \dfrac{1}{6} \cos 3x$

4 $\dfrac{1}{4}$

5 $\dfrac{1}{2}(1 - \ln 2)$

Practice questions I

1 (a) $\dfrac{20}{3}$ (b) $\dfrac{62}{5}$ (c) $\dfrac{8\pi}{3} - 2\sqrt{3}$

  (d) $\dfrac{5}{12}$

2 (a) $3y + 2x - 5 = 0$ (b) $\dfrac{23}{20}$

3 (a) $t = 0, t = \dfrac{\pi}{2}$ (b) $12 c^2 \pi$

4

108.8

5 (a) $y = 2x - 8$ (c) 16.2

SECTION 6

First-order differential equations

INTRODUCTION With our greatly increased range of integration techniques we can start to explore the methods of solution of differential equations beyond the very basic type we looked at in P1. Fundamental to this is the idea of the rate of change of one variable with respect to another: the second variable is usually time but can also be distance, for example. We look particularly at one kind of equation, the solution of which represents exponential growth or decay. This is a topic that students frequently find difficult so there are a large number of questions for you to practise the various techniques involved.

Separable variables

OCR P3 5.3.6 (b)

In P1 we saw that differentiating an explicit function, say $y = 4x^2 - 3x + 5$, gives a gradient function where only x's appear: in this case

$$\frac{dy}{dx} = 8x - 3$$

Then we saw that we can reverse this by integrating and in fact can recapture the exact original function provided that we know a particular point that the curve passes through, e.g. (1, 6)

$$y = \int (8x - 3)\, dx = 4x^2 - 3x + C \qquad \text{General}$$

Putting in the point, $6 = 4 - 3 + C \;\Rightarrow\; C = 5$

giving the original equation, $y = 4x^2 - 3x + 5$. When we differentiate an implicit function, like $x^2 + y^2 = 4x$, we have a different situation

$$2x + 2y\frac{dy}{dx} = 4 \;\Rightarrow\; 2y\frac{dy}{dx} = 4 - 2x$$

$$\frac{dy}{dx} = \frac{4 - 2x}{2y} = \frac{2 - x}{y}$$

Now we cannot simply integrate the RHS for y: there is a mixture of variables x and y. With implicit functions, the equation connecting $\frac{dy}{dx}$ with variables x and y can be more or less complicated; for the moment we shall be looking at a particular kind where we can separate the variables by division and multiplication.

This kind of equation is called, reasonably enough, separable variables and the idea is to end up with all the y's on the LHS and all the x's on the RHS so that we have $g(y)\frac{dy}{dx} = f(x)$

where f and g are the respective functions of x and y.

85

P3 Section 6

Examples

(a) $y \dfrac{dy}{dx} = x^2 (1 + y)$

Dividing both sides by $(1 + y)$ gives

$$\left(\dfrac{y}{1+y}\right) \dfrac{dy}{dx} = x^2$$

and this is in the correct form

(b) $\sin y \dfrac{dy}{dx} + x^2 = x^2 e^y$

This doesn't look as though it's suitable but in fact it is. After rearranging, we get:

$$\sin y \dfrac{dy}{dx} = x^2 e^y - x^2$$

$$= x^2 (e^y - 1) \qquad \text{factorising}$$

$$\dfrac{\sin y}{e^y - 1} \dfrac{dy}{dx} = x^2 \qquad \text{dividing by } (e^y - 1)$$

On the other hand, something like:

(c) $\dfrac{dy}{dx} = x + y$

is not of this type. The addition sign means that we cannot separate the x and y.

Be careful with powers, they can be misleading. Given the equation:

(d) $e^{2x+y} \dfrac{dy}{dx} = x^2$

you would think that the $2x$ can't be separated from the y, which is not true. Remembering that powers are added, we can write

$$e^{2x+y} = e^{2x} \times e^y$$

and the equation becomes

$$e^{2x} e^y \dfrac{dy}{dx} = x^2$$

$$e^y \dfrac{dy}{dx} = \dfrac{x^2}{e^{2x}} = x^2 e^{-2x}$$

Solving the equations

> OCR P3 5.3.6 (b),(c)

Once we have rearranged the equation into the correct form, we have to have a look at ways of solving them, that is, to find what the function was before differentiation.

When we have something like:

$$2y \dfrac{dy}{dx} = \sin x$$

we have to try and work out where the left-hand side came from.

Which function gives $2y \dfrac{dy}{dx}$ after differentiation with respect to x?

To find out, we try to integrate both sides with respect to x, and the equation becomes

$$\int 2y \dfrac{dy}{dx} dx = \int \sin x \, dx$$

But $\dfrac{dy}{dx} dx$ is equivalent to dy, as we saw in the section on substitution, so we have

$$\int 2y \, dy = \int \sin x \, dx$$

Each side can now be treated in the same manner, just as two normal functions to be integrated, and then we would have:

$$y^2 = -\cos x + C$$

Note that we only need one arbitrary constant even though we have two different integrations. Since the constants on each side are unknown, we can combine them to form a single unknown constant. In fact, together with the differential equation we are usually given limits, a pair of corresponding values for the variable, and from these we can work out the value of this constant. Suppose in the above example that $y = 2$ when $x = 0$. We put these values into our general solution and then:

$$2^2 = -\cos 0 + C$$
$$4 = -1 + C \Rightarrow C = 5$$

Having found C, we can rewrite our solution as:

$$y^2 = 5 - \cos x$$

We will have a look at two more examples, and then you will be able to try some.

Example

Find the general solution for the equation

$$\dfrac{dy}{dx} + 2xy = 2x$$

in the form $y = f(x)$.

Solution

Rearranging, $\dfrac{dy}{dx} = 2x - 2xy$

$= 2x(1 - y)$

$\Rightarrow \dfrac{1}{1-y} \dfrac{dy}{dx} = 2x$

i.e. $\int \dfrac{1}{1-y} dy = \int 2x \, dx$

$\Rightarrow -\ln(1-y) = x^2 + C$

$\ln(1-y) = -x^2 - C$

Taking e's, $1 - y = e^{-x^2 - C}$ (NB, not $e^{-x^2} - e^C$!)

$= e^{-x^2} \times e^{-C} = A e^{-x^2}$ where $A = e^{-C}$

Finally, $y = 1 - A e^{-x^2}$, the required form.

P3 Section 6

The procedure whereby the unknown constant e^{-C} is changed into another unknown constant A is quite common.

Example

The differential equation:

$$x \frac{dy}{dx} + \tan y = 0$$

where $x > 0$ and $0 < y < \frac{\pi}{2}$, satisfies the condition $y = \frac{1}{3}\pi$ when $x = 2$.

Show that the solution may be expressed in the form $x \sin y = k$, where k is a constant whose value is to be stated.

Solution

We'll rearrange the equation first of all:

$$x \frac{dy}{dx} = -\tan y$$

And again: $\quad \dfrac{-1}{\tan y} \dfrac{dy}{dx} = \dfrac{1}{x}$

Integrate both sides with respect to x:

$$\int \frac{-1}{\tan y} \frac{dy}{dx} dx = \int \frac{1}{x} dx$$

$$\int \frac{-\cos y}{\sin y} dy = \ln x$$

$$-\ln \sin y = \ln x + C \qquad \qquad \ldots \text{①}$$

Now we'll put in the limits that are given, $y = \dfrac{\pi}{3}$ when $x = 2$

$$-\ln \sin \frac{\pi}{3} = \ln 2 + C$$

$$-\ln \frac{\sqrt{3}}{2} = \ln 2 + C$$

$$C = -\ln \frac{\sqrt{3}}{2} - \ln 2$$

$$= -\left[\ln \frac{\sqrt{3}}{2} + \ln 2\right] = -\left[\ln \sqrt{3}\right]$$

Putting our value for C back in the solution ①, we get:

$$-\ln \sin y = \ln x - \ln \sqrt{3}$$

If we take $-\ln \sin y$ to the other side and combine, i.e.:

$$0 = \ln \frac{x \sin y}{\sqrt{3}}$$

$$\frac{x \sin y}{\sqrt{3}} = 1$$

then: $\quad x \sin y = \sqrt{3}$

which is in the form required, with $k = \sqrt{3}$.

Notice the fact that both the integrals involved ln – this is common.

First-order differential equations

Practice questions A

1. Find general solutions for the following, giving your answer in the form $y = f(x)$

   (a) $\dfrac{1}{y}\dfrac{dy}{dx} = 2x$ (b) $\dfrac{x}{y}\dfrac{dy}{dx} + 1 = 0$ (c) $\dfrac{dy}{dx} - \dfrac{y}{x} = 0$ (d) $\dfrac{1}{x}\dfrac{dy}{dx} = \dfrac{1}{2y^3}$

   (e) $\dfrac{dy}{dx} = (2 \sin x)\, y$ (f) $\dfrac{dy}{dx} + 3y = 0$

2. Find particular solutions for the following:

   (a) $\dfrac{dy}{dx} = \dfrac{x}{y}$: $x = 2$ when $y = 1$ (b) $\dfrac{dy}{dx} = -xy^2$: $y = -\dfrac{1}{2}$ when $x = 2$

   (c) $\dfrac{dy}{dx} = 3x^2 y$: $y = 2$ when $x = 0$ (d) $2x + 4y\dfrac{dy}{dx} = 1$: $y = 1$ when $x = 0$

   (e) $x\dfrac{dy}{dx} = (1 - 2x^2)\, y$: $y = 1$ when $x = 1$

3. If $\dfrac{dy}{dx} = \dfrac{-y}{x^2}$ and $y = e$ when $x = 1$, show that $y = e^{\frac{1}{x}}$

Partial fractions

Questions frequently involve partial fractions: in this case you have to rearrange lns and fractions before ending up with a solution in the form asked for.

Example

Express $\dfrac{x}{(x+1)(x+2)}$ in partial fractions.

Solve the differential equation:
$$(x+1)(x+2)\dfrac{dy}{dx} = x(y+1)$$
for $x > -1$, given that $y = \dfrac{1}{2}$ when $x = 1$. Express your answer in the form $y = f(x)$.

Solution

Note that the restriction on the values for x is to ensure that the denominator of the fraction:
$$\dfrac{x}{(x+1)(x+2)}$$
is never 0, which otherwise could lead to some strange results. Using partial fractions, we get that:
$$x \equiv A(x+2) + B(x+1)$$
$x = -2$ gives $B = 2$ and $x = -1$ gives $A = -1$, so that:
$$\dfrac{x}{(x+1)(x+2)} = \dfrac{2}{x+2} - \dfrac{1}{x+1} \qquad \ldots \text{①}$$

Rearranging the differential equation, we get:
$$\dfrac{1}{(y+1)}\dfrac{dy}{dx} = \dfrac{x}{(x+1)(x+2)}$$

Integrating both sides with respect to x gives:
$$\int \dfrac{1}{y+1}\dfrac{dy}{dx}\, dx = \int \dfrac{x\, dx}{(x+1)(x+2)} = \int \left(\dfrac{2}{x+2} - \dfrac{1}{x+1}\right) dx \qquad \text{using ①}$$

Integrate:
$$\ln(y+1) = 2\ln(x+2) - \ln(x+1) + C \qquad \ldots ②$$

Put in the limits of $y = \frac{1}{2}$ when $x = 1$:

$$\ln\frac{3}{2} = 2\ln 3 - \ln 2 + C$$
$$\ln\frac{3}{2} = \ln\frac{9}{2} + C$$
$$C = \ln\frac{3}{2} - \ln\frac{9}{2} = \ln\frac{1}{3}$$

and put this back into [2]:

$$\ln(y+1) = 2\ln(x+2) - \ln(x+1) + \ln\frac{1}{3}$$
$$\ln(y+1) = \ln\frac{(x+2)^2}{3(x+1)}$$

i.e.: $y + 1 = \dfrac{(x+2)^2}{3(x+1)}$

$$y = \frac{(x+2)^2}{3(x+1)} - 1$$

Then: $y = \dfrac{x^2 + x + 1}{3(x+1)}$

when we put both the terms together as one fraction.

Isolating y

In this example there was very little rearranging to present our solution for y in terms of x. It can happen that y occurs at the top and bottom of a fraction, something like:

$$\frac{y+2}{y-1} = x - 3$$

To isolate y, we first of all cross-multiply to clear all fractions:

$$y + 2 = (y-1)(x-3)$$
$$= yx - x - 3y + 3$$

Then we gather all terms containing y to one side, and all terms without y to the other:

$$y - yx + 3y = -x + 3 - 2$$
$$4y - yx = 1 - x$$

Then we can factorise the left-hand side,

$$y(4 - x) = 1 - x$$

and $\quad y = \dfrac{1-x}{4-x}$

Practice questions B

1 Solve the differential equations:

(a) $(x-1)(2-x)\dfrac{dy}{dx} = y : y = 1$ when $x = \dfrac{3}{2}$

(b) $\dfrac{dy}{dx} = y(1-y) : y = \dfrac{1}{2}$ when $x = 0$

(c) $\cos\theta \dfrac{dr}{d\theta} + r\sin\theta = 0 : r = a$ when $\theta = 0$

(d) $x\dfrac{dy}{dx} + y(1+x^2) = 0 : y = e^{-\frac{1}{2}}$ when $x = 1$

(e) $x^2(x-1)\dfrac{dy}{dx} = y : y = e^{\frac{1}{2}}$ when $x = 2$

(f) $2x(2x-1)\dfrac{dy}{dx} + 1 = y^2 : y = \dfrac{3}{2}$ when $x = \dfrac{5}{9}$

Rates of change

OCR P3 5.3.6 (a)

There is one further topic that we have to cover while we're here. Occasionally, the differential equation is not given and you are expected to form it yourself from the data given. It very often involves rates of change of one or more variable, which in effect is the variable differentiated with respect to time. For example, the rate of change of a volume V is expressed $\dfrac{dV}{dt}$, or the rate of change of a radius r is written $\dfrac{dr}{dt}$.

There are two basic facts that you will need to set up a differential equation:

1 If the rate of change is increasing, it is positive, whilst a decrease means it is negative.

2 If a variable is proportional to some other variable, we express it as $k \times$ other variable, and use the limits given to evaluate k.

Let's have a look at a couple of examples of these.

Example

Express the following information in the form of a differential equation.

(a) The surface of a pond is partially covered with weed. The weed is increasing in area at a rate proportional to its area at that instant. (Call the area of the weed x m^2 and the growing time, t days).

(b) The rate at which the temperature, $\theta°$, of a body is decreasing is proportional to the difference between θ and the constant temperature of the medium, $\overset{\circ}{\theta}_m$

Solution

(a) The rate of change of the area of the weed is written $\dfrac{dx}{dt}$ and it's positive since it's increasing.

The differential equation is: $\dfrac{dx}{dt} = kx$

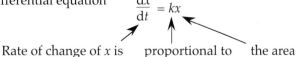

Rate of change of x is   proportional to   the area

(b) The rate of change of temperature, $\dfrac{d\theta}{dt}$, is negative since it's decreasing

i.e. $\dfrac{d\theta}{dt} = -k(\theta - \theta_m)$

Note that in both cases we used a constant of proportionality, k, which conventionally is taken to be positive.

Both of these equations are examples of exponential growth and decay, respectively, which we had a brief look at in the exponential section in P2. As was mentioned there, the solution of this type of equation has many practical uses. We'll have a look at some typical examples.

Example

A solution contains x bacteria at time t hours, and it is known that x increases at a rate proportional to x.

At $t = 0$, $x = 1000$; at $t = 2$, $x = 9000$. Find x when $t = 1$ and find an expression for x in terms of t.

Solution

The rate of increase of x is proportional to x: in symbols this is

$$\dfrac{dx}{dt} = kx$$

$$\dfrac{1}{x}\dfrac{dx}{dt} = k$$

$$\Rightarrow \int \dfrac{1}{x} dx = \int k\, dt$$

$\ln |x| = kt + C$ ... ①

We have two unknown constants, k and C, so we need two pairs of values of the variables x and t.

When $t = 0$, $x = 1000$:   these into ① give

$\ln 1000 = C$ and substituting this back into ①

$\ln |x| = kt + \ln 1000$ ... ②

When $t = 2$, $x = 9000$: these into ② give

$\ln 9000 = 2k + \ln 1000$

$\ln 9000 - \ln 1000 = 2k$

$\ln 9 = 2k \Rightarrow k = \dfrac{1}{2}\ln 9 = \ln 3$

Back into ② gives

$\ln |x| = t \ln 3 + \ln 1000$ ... ③

When $t = 1$, $\ln x = \ln 3 + \ln 1000 = \ln 3000$

$\Rightarrow x = 3000$

The term $t \ln 3$ in ③ is the same as $\ln 3^t$, so we can rearrange ③ to

$\ln x \quad = \ln 3^t + \ln 1000$

$\quad\quad\quad = \ln 1000 \cdot 3^t$

$\Rightarrow x \quad = 1000 \cdot 3^t$

This is a typical solution of an exponential growth equation, of the form

$x = a \times b^t$ where a and b are constants.

a is the value of x when $t = 0$: this is frequently written x_0 or equivalent. b is the factor increase in unit time – it is like r, the common ratio in a GP.

The graph of x against t looks like this:

Figure 6.1

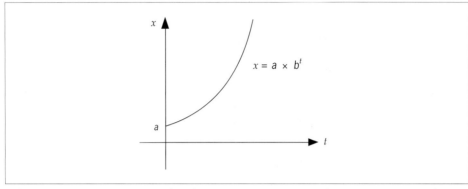

This is exponential *growth*: the characteristic feature is that it increases by the same factor during two successive time intervals. For example, if it doubles from 80 to 160 in a certain time, then it will double from 160 to 320 in another interval of the same length.

Exponential decay has the same feature, but the ratio is less than 1, i.e. it is decreasing. The graph of a typical exponential decay looks like this:

Figure 6.2

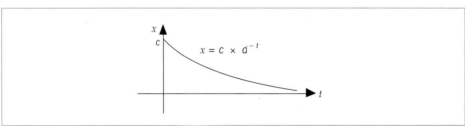

Here's an example of this type.

Example

In a chemical reaction, hydrogen peroxide is converted into water and oxygen. At time t after the start of the reaction, the quantity of hydrogen peroxide that has not been converted is h and the rate at which h is decreasing is proportional to h. Write down a differential equation involving h and t. Given that $h = H$ initially, show that

$$\ln \frac{h}{H} = -kt$$

where k is a positive constant.

In an experiment, the time taken for the hydrogen peroxide to be reduced to half of its original quantity was 3 minutes. Find, to the nearest minute, the time that would be required to reduce the hydrogen peroxide to one-tenth of its original quantity.

Express h in terms of H and t, [and sketch a graph showing how h varies with t.]

Section 6

Solution

Here, the rate of decrease of h is proportional to h, i.e.

$$\frac{dh}{dt} = -kh$$

$$\frac{1}{h}\frac{dh}{dt} = -k$$

$$\Rightarrow \int \frac{1}{h}\,dh = \int -k\,dt$$

$$\ln h = -kt + C$$

when $t = 0$, $h = H$ $\Rightarrow \ln H = C$

$$\ln h = -kt + \ln H$$

$$\ln h - \ln H = -kt$$

$$\Rightarrow \ln \frac{h}{H} = -kt \quad \ldots \text{①} \text{ as required.}$$

We are given that when $h = \frac{H}{2}$, $t = 3$.

Putting these in, $\ln \frac{1}{2} = -3k$

$$\Rightarrow k = \frac{-1}{3}\ln \frac{1}{2} = \frac{1}{3}\ln 2$$

i.e. $\ln \frac{h}{H} = \frac{-1}{3}t \ln 2 \quad \ldots \text{②}$

When $h = \frac{H}{10}$, $\ln \frac{1}{10} = \frac{-1}{3}t \ln 2$

$$\Rightarrow t = \frac{-3\ln \frac{1}{10}}{\ln 2} = 10 \text{ mins (nearest)}$$

The RHS of ② is the same as $\ln 2^{\frac{-t}{3}}$

and so ② becomes $\ln \frac{h}{H} = \ln 2^{\frac{-t}{3}}$

$$\frac{h}{H} = 2^{\frac{-t}{3}}$$

$$h = H \times 2^{\frac{-t}{3}}$$

[Note that if we write $2^{\frac{-t}{3}}$ as $\left(2^{\frac{1}{3}}\right)^{-t}$, we have the standard form for exponential decay, i.e. $x = x_0 \times A^{-t}$.]

The example that follows is quite a standard type of exponential decay – a hot body cooling in colder surroundings according to Newton's law of cooling.

Example

(a) Newton's law of cooling states that the rate at which an object cools is proportional to the difference in temperature between the object and its surroundings.

The temperature, $\theta°C$, of a hot drink, t minutes after it has been poured, satisfies the differential equation

$$\frac{d\theta}{dt} = -a(\theta - b)$$

where a and b are constants. The temperature of the surroundings of the drink is 25°C.

Write down the value of b.

The rate of cooling when $\theta = 65$ is 8°C per minute. Find the value of a.

(b) The temperature, ϕ°C, of another hot drink, t minutes after being poured, satisfies the differential equation

$$\frac{d\phi}{dt} = -k(\phi - 20)$$

where k is a constant.

(i) Solve this differential equation to show that $\phi = A + Be^{-kt}$, where A and B are constants and the value of A is to be found

(ii) Given that $\phi = 80$ when $t = 0$ and that $\phi = 50$ when $t = 2$, find the value of B and k.

(c) An object in an industrial oven has temperature T°C at time t, where

$$T = 1000 + 200e^{\sin t}$$

and t is measured in days.

Find a function $f(t)$ and a constant c such that

$$\frac{dT}{dt} = f(t)(T - c)$$

Solution

(a) $b = 25$

When $\theta = 65$, $\dfrac{d\theta}{dt} = -8$ $\quad \Rightarrow -8 = -a(65 - 25)$

$\Rightarrow a = \dfrac{1}{5}$

(b) (i) Rearranging, $\displaystyle\int \frac{1}{\phi - 20} \, d\phi = \int -k \, dt$

$\ln(\phi - 20) = -kt + C$

Taking e's, $\phi - 20 = e^{-kt + C}$

$\phi = 20 + Be^{-kt}$ ... ① $(B = e^C)$

and $A = 20$

(ii) Putting $\phi = 80$ and $t = 0$ into ①, $80 = 20 + B \Rightarrow B = 60$

$\phi = 20 + 60e^{-kt}$ ... ②

$\phi = 50$ and $t = 2$ into ②, $50 = 20 + 60e^{-2k}$

$e^{-2k} = \dfrac{1}{2}$

$-2k = \ln \dfrac{1}{2}$

$k = -\dfrac{1}{2} \ln \dfrac{1}{2} = \dfrac{1}{2} \ln 2$

i.e. $B = 60$ and $k = \dfrac{1}{2} \ln 2$

(c) Differentiating, $\dfrac{dT}{dt} = 200 \cos t\, e^{\sin t}$

$\qquad\qquad\qquad = \cos t\, (200\, e^{\sin t})$

$\qquad\qquad\qquad = \cos t\, (T - 1000)$

and $f(t) = \cos t$, $C = 1000$

Related rates of change

OCR P3 5.3.6

Questions can sometimes involve two related variables, e.g. volume and depth of water, and the information given relates one rate of change with the other variable, e.g. the rate of decrease of the volume is proportional to the square root of the depth. This leads to the differential equation

$$\dfrac{dV}{dt} = -k\sqrt{x} \qquad \qquad \ldots \text{①}$$

and we can't solve this as it stands: we have to reduce the number of variables.

We need to find an equation connecting the two variables: in this case, if we suppose that there is constant cross section and the base area is 10 m², then the equation is

$\qquad$ Volume $=$ area × height

i.e. $V = 10x$

When we differentiate this we have $\dfrac{dV}{dx}$, which serves as a link between $\dfrac{dV}{dt}$

and $\dfrac{dx}{dt}$, using the chain rule.

$\qquad \dfrac{dV}{dx} = 10$ and $\dfrac{dV}{dt} = \dfrac{dV}{dx} \times \dfrac{dx}{dt}$

i.e. $\dfrac{dV}{dt} = 10 \dfrac{dx}{dt}$

The equation ① now becomes

$\qquad 10 \dfrac{dx}{dt} = -k\sqrt{x}$

and we can solve this as normal.

Example

In an irrigation system, water is stored in a rectangular tank with a square horizontal base of edge 200 cm and with a vertical height of 400 cm. A tap in the base is opened and water flows out. After t seconds, when the depth of water in the tank is x cm, the rate of flow is $100\sqrt{x}$ cm³ s⁻¹.

Show that

$\qquad 400 \dfrac{dx}{dt} + \sqrt{x} = 0$

The tank is initially full. When the tap is opened, find to the nearest half hour how long the tank takes to empty.

Also find the depth of water in the tank after the tap has been open for 2 hours.

Solution

We have two related variables which are changing with respect to time, the volume V and the depth of water x: their rates of change are $\dfrac{dV}{dt}$ and $\dfrac{dx}{dt}$ respectively. The area of the base is 200 cm × 200 cm and so V and x are connected via the equation

$$V = 200^2 x \implies \frac{dV}{dx} = 200^2 \qquad \ldots \text{①}$$

'Rate of flow' refers to the change in volume, $\dfrac{dV}{dt}$,

$$\frac{dV}{dt} = -100\sqrt{x} \qquad \ldots \text{②}$$

Using the chain rule, $\dfrac{dV}{dt} = \dfrac{dV}{dx} \times \dfrac{dx}{dt}$

$$= 200^2 \times \frac{dx}{dt} \quad \text{(from ①)}$$

This into ② gives $200^2 \times \dfrac{dx}{dt} = -100\sqrt{x}$

$$\implies 400 \frac{dx}{dt} = -\sqrt{x} \qquad \ldots \text{③}$$

or $400 \dfrac{dx}{dt} + \sqrt{x} = 0$ as required

Rearranging ③, $\displaystyle\int \frac{400}{\sqrt{x}} dx = \int -dt$

$$\implies 800\sqrt{x} = -t + C$$

'Initially full' means when $t = 0$, $x = 400$

$$\implies 800\sqrt{400} = 0 + C \implies C = 16\,000$$

and $800\sqrt{x} = 16\,000 - t$ ... ④

When empty, $x = 0$ and so

$$0 = 16000 - t \implies t = 16\,000 \text{ sec.}$$
$$= 4\tfrac{1}{2} \text{ hours (nearest } \tfrac{1}{2} \text{ hr)}$$

When $t = 2$ hours $= 7200$.

Into ④, $800\sqrt{x} = 16\,000 - 7200 = 8800$

$$\implies \sqrt{x} = 11 \text{ and } x = 121$$

Practice questions C

1 Find a relationship between the given rates of change by differentiating the given equation relating the variables and applying the chain-rule.

(a) $\dfrac{dA}{dt}$ and $\dfrac{dr}{dt}$ when $A = \pi r^2$

(b) $\dfrac{dV}{dt}$ and $\dfrac{dh}{dt}$ when $V = 100h$

(c) $\dfrac{dV}{dt}$ and $\dfrac{dr}{dt}$ when $V = \dfrac{4}{3}\pi r^3$

(d) $\dfrac{dS}{dt}$ and $\dfrac{dr}{dt}$ when $S = 4\pi r^2$

P3 Section 6

2 Using the formulae in (c) and (d) above, find an expression for $\frac{dV}{dt}$ in terms of S and $\frac{dS}{dt}$ by using the extended chain-rule.

$$\frac{dV}{dt} = \frac{dV}{dr} \times \frac{dr}{dS} \times \frac{dS}{dt}$$

More than one factor

OCR P3 5.3.6

There may be more than one factor affecting the rate of change of a variable: one amount being added and another taken away, for example. This makes the equation a little more complicated, but so long as we are careful with the rearrangement, it doesn't necessarily make the working much more difficult. Remember that a factor is positive if increasing the variable, negative if decreasing.

Example

A much-simplified model for the filling of a new reservoir is as follows. Initially, the reservoir is empty. At time t years after starting to be filled, the volume of water it contains is V cubic metres and the depth of water is h metres, where V is proportional to h. The rate at which water enters the reservoir is constant, but there is a loss of water (due to leakage and evaporation), the rate of which is proportional to h.

Show that this model leads to the differential equation

$$\frac{dh}{dt} = \lambda - \mu h$$

where λ and μ are constants.

Given that $\lambda = 300$ and $\mu = 1.5$, solve this equation for h as a function of t.

It is possible to start drawing water from the reservoir when the depth reaches 150 metres. Calculate the value of t when this occurs.

Solution

We are given that V is proportional to h, i.e. that $V = k_1 h$ for some constant k_1

then $\frac{dV}{dh} = k_1$ ... ①

'The rate at which water enters' refers to the volume, so we have the positive factor for $\frac{dV}{dt}$ of R, where R is also a constant.

The leakage, which is a negative factor for $\frac{dV}{dt}$, is $k_2 h$. Overall we have,

$$\frac{dV}{dt} = R - k_2 h$$

Chain-rule: $\frac{dV}{dt} = \frac{dV}{dh} \times \frac{dh}{dt} = k_1 \times \frac{dh}{dt}$ from ①

i.e. $k_1 \frac{dh}{dt} = R - k_2 h$

$\Rightarrow \frac{dh}{dt} = \frac{R}{k_1} - \frac{k_2}{k_1} h$

$= \lambda - \mu h$

where $\lambda = \frac{R}{k_1}$ and $\mu = \frac{k_2}{k_1}$.

If $\lambda = 300$ and $\mu = 1.5$

$$\frac{dh}{dt} = 300 - 1.5h$$

Rearranging, $\int \frac{1}{300 - 1.5h} \, dh = \int dt$

$$-\frac{1}{1.5} \ln |300 - 1.5h| = t + C \qquad \ldots \text{②}$$

'Initially empty' $\Rightarrow t = 0, h = 0$

$$-\frac{1}{1.5} \ln 300 = C$$

into ②, $\frac{1}{1.5} \left[\ln 300 - \ln(300 - 1.5h) \right] = t$

$$\frac{1}{1.5} \left[\ln \frac{300}{300 - 1.5h} \right] = t$$

$$\ln \frac{300}{300 - 1.5h} = 1.5t \qquad \ldots \text{③}$$

$$\frac{300}{300 - 1.5h} = e^{1.5t}$$

$$300 \, e^{-1.5t} = 300 - 1.5h$$

$$1.5h = 300(1 - e^{-1.5t})$$

$$\Rightarrow h = 200(1 - e^{-1.5t})$$

When $h = 150$, into ③ gives

$$\ln \frac{300}{300 - 225} = 1.5t$$

$$\ln 4 = 1.5t$$

$$\Rightarrow t = \frac{2}{3} \ln 4 \text{ (years)}$$

SUMMARY EXERCISES

1 Find the general solution of the differential equation

$$\frac{dy}{dx} - \frac{2y}{x} = 0$$

Sketch the two solution curves passing respectively through the points $(2, 2)$ and $(-1, -1)$.

2 Solve the differential equation

$$2x^2 y \frac{dy}{dx} = (1 + x^2)(1 + y^2)$$

given that $y = 1$ when $x = 1$. Give your answer in the form $y^2 = f(x)$.

3 In the x-y plane, O is the origin, A is the point $(2, 0)$ and P is a general point (x, y). Write down the gradients of OP and AP in terms of x and y.

A curve has the property that the gradient of the normal to the curve at the general point $P(x, y)$ is equal to the product of the gradients of OP and AP, where O and A are the points defined above. Use this information to show that

$$\frac{dy}{dx} = \frac{2x - x^2}{y^2}$$

Given that the curve passes through the point $(1, 1)$, find its equation.

4 (a) Find $\int x e^{-x} \, dx$

(b) Given that $y = \frac{\pi}{4}$ at $x = 0$, solve the differential equation

$$e^x \frac{dy}{dx} = \frac{x}{\sin 2y}$$

5 The radius, r cm, of a circular oil patch is increasing, at time t seconds, at a rate which is proportional to $1/r$. When the radius is 2 cm it is increasing at a rate of 0.1 cm s^{-1}. Show that r satisfies the differential equation

$$r \frac{dr}{dt} = 0.2$$

Find the time taken for the area of the oil patch to increase from 4π cm^2 to 16π cm^2.

6 (a) Using the substitution $u = 1 + 2x$, or otherwise, find

$$\int \frac{4x}{(1+2x)^2} \, dx, \quad x > -\frac{1}{2}$$

(b) Given that $y = \frac{\pi}{4}$ when $x = 0$, solve the differential equation

$$(1+2x)^2 \frac{dy}{dx} = \frac{x}{\sin^2 y}$$

7 The rate in cm^3 s^{-1}, at which oil is leaking from an engine sump at any time t seconds is proportional to the volume of oil, V cm^3, in the sump at that instant. At time $t = 0$, $V = A$.

(a) By forming and integrating a differential equation, show that

$$V = Ae^{-kt},$$

where k is a positive constant.

(b) Sketch a graph to show the relation between V and t.

Given further that $V = \frac{1}{2}A$ at $t = T$,

(c) show that $kT = \ln 2$.

8 A biologist studying fluctuations in the size of a particular population decides to investigate a model for which

$$\frac{dP}{dt} = kP \cos kt,$$

where P is the size of the population at time t days and k is a positive constant.

(a) Given that $P = P_0$ when $t = 0$, express P in terms of k, t and P_0.

(b) Find the ratio of the maximum size of the population to its minimum size.

9 At time t hours the rate of decay of the mass of a radioactive substance is proportional to the mass x kg of the substance at that time. At time $t = 0$, the mass of the substance is A kg.

(a) By forming and integrating a differential equation, show that $x = Ae^{-kt}$, where k is a constant.

It is observed that $x = \frac{1}{3}A$ at time $t = 10$.

(b) Find the value of t when $x = \frac{1}{2}A$, giving your answer to 2 decimal places.

10 In a certain pond, the rate of increase of the number of fish is proportional to the number of fish, n, present at time t. Assuming that n can be regarded as a continuous variable, write down a differential equation relating n and t, and hence show that

$$n = Ae^{kt},$$

where A and k are constants.

In a revised model, it is assumed also that fish are removed from the pond, by anglers and by natural wastage, at the constant rate of p per unit time, so that

$$\frac{dn}{dt} = kn - p.$$

Given that $k = 2$, $p = 100$ and that initially there were 500 fish in the pond, solve this differential equation, expressing n in terms of t.

Give a reason why this revised model is not satisfactory for large values of t.

11 A water tank is cuboid in shape, with rectangular horizontal cross section of area 8 m^2 and depth 1 m. At time $t = 0$ the tank is empty. Water is poured into the tank at a constant rate of 0.01 m^3 s^{-1} and leaks from a hole in the base at a rate of $\frac{1}{4}x^2$ m^3 s^{-1}, where x is the depth of the water in metres at time t seconds.

(a) Show that $\frac{dx}{dt} = \lambda - \mu x^2$, giving the values of the constants λ and μ.

(b) Find the time taken for the water in the tank to reach a depth of 0.1 m.

12 (a) Express $\dfrac{1}{(3x-1)x}$ in partial fractions.

A model for the way in which a population of animals in a closed environment varies with time is given, for $P > \tfrac{1}{3}$, by

$$\dfrac{dP}{dt} = \tfrac{1}{2}(3P^2 - P)\sin t$$

where P is the size of the population in thousands at time t.

(b) Given that $P = \tfrac{1}{2}$ when $t = 0$, use the method of separation of variables to show that

$$\ln\left(\dfrac{3P-1}{P}\right) = \tfrac{1}{2}(1 - \cos t).$$

(c) Calculate the smallest positive value of t for which $P = 1$.

(d) Rearrange the equation at the end of part (b) to show that

$$P = \dfrac{1}{3 - e^{\tfrac{1}{2}(1-\cos t)}}.$$

Hence find the two values between which the number of animals in the population oscillates.

13 In a simple model of population growth, the rate of increase of the population at any time is taken to be proportional to the population at that time. Treating the size of the population as a continuous variable, this model leads to the differential equation

$$\dfrac{dP}{dt} = kP,$$

where P denotes the size of the population at time t (measured in days), and k is a positive constant.

(a) Given that $P = P_0$ when $t = 0$, express P in terms of k, t and P_0.

(b) In a laboratory experiment, the number of individual cells present in a specimen was 1218 at a certain time, and was 1397 one day later. Find the number of cells after one further day predicted by the model.

(c) Given a reason why this model can never represent a real-world population correctly over the longer term.

Biologists studying a particular population observe that the fertility of individual members in the population appears to be decreasing as time goes on, and that consequently the population is not growing as rapidly as had been expected. They propose that the situation may be modelled mathematically by replacing the constant k in the simple model above by a term of the form $Ke^{-\lambda t}$, where K and λ are positive constants. This revised model leads to the differential equation

$$\dfrac{dP}{dt} = Ke^{-\lambda t} P,$$

where $P = P_0$ when $t = 0$. Show that this model predicts a population that approaches a certain fixed limit, and express this limit in terms of K, λ and P_0.

14 When a casserole was removed from an oven at 6 pm, its temperature was 150°C. Its temperature had fallen to 70°C by 6.20 pm. The casserole can be put into the freezer when its temperature has dropped to 30°C.

(a) Estimate when this will be by assuming a constant rate of cooling from 150°C.

(b) An improved model of the casserole's temperature, θ°C, at time t minutes after 6 pm, can be deduced from the assumptions that:

the rate of loss of temperature is proportional to the difference in temperature between the casserole and the kitchen;

the kitchen remains at a constant 20°C.

Show that these assumptions lead to an equation of the form

$$t = a \ln(\theta - 20) + b,$$

where a and b are constants.

Find a and b, and hence obtain a new estimate for the time at which the casserole can be put into the freezer.

15 A rectangular water tank has a horizontal square base of side 1 metre. Water is flowing out of the tank from an outlet in the base but is also being pumped into the tank at a constant rate of 400 cm^3 s^{-1}.

Initially, the depth of water in the tank is 81 cm and water is flowing out at 900 cm^3 s^{-1}.

(a) Assuming that the rate at which the water flows out remains constant, calculate the time, in seconds, taken for the depth of water to decrease from 81 cm to 64 cm.

(b) In an improved model, it is assumed that water flows from the outlet at a rate which is proportional to the square root of the depth, h cm, of water in the tank.

 (i) Show how this model leads to the differential equation
$$\frac{dh}{dt} = 0.04 - 0.01\sqrt{h},$$
where t is the time in seconds.

 (ii) Show that the solution of the differential equation in (i) is given by
$$t = \int \frac{100}{4 - \sqrt{h}}\, dh.$$
Use the substitution $x = \sqrt{h} - 4$ to find the time, in seconds, taken for the depth of water to decrease from 81 cm to 64 cm.

16 A metal rod is 60 cm long and is heated at one end. The temperature at a point on the rod at distance x cm from the heated end is denoted by $T\,°C$. At a point half way along the rod, $T = 290$ and $\frac{dT}{dx} = -6$.

(a) In a simple model for the temperature of the rod, it is assumed that $\frac{dT}{dx}$ has the same value at all points on the rod. For this model, express T in terms of x and hence determine the temperature difference between the ends of the rod.

(b) In a more refined model, the rate of change of T with respect to x is taken to be proportional to x. Set up a differential equation for T, involving a constant of proportionality k.

Solve the differential equation and hence show that, in this refined model, the temperature along the rod is predicted to vary from 380°C to 20°C.

17 The area of a circle of radius r metres is A m^2.

(a) Find $\frac{dA}{dr}$ and write down an expression, in terms of r, for $\frac{dr}{dA}$.

(b) The area increases with time t seconds in such a way that
$$\frac{dA}{dt} = \frac{2}{(t+1)^3}.$$
Find an expression, in terms of r and t, for $\frac{dr}{dt}$.

(c) Solve the differential equation
$$\frac{dA}{dt} = \frac{2}{(t+1)^3}$$
to obtain A in terms of t, given that $A = 0$ when $t = 0$.

(d) Show that, when $t = 1$, $\frac{dr}{dt} = 0.081$, correct to 2 significant figures.

18 The initial volume of a spherical mothball is 5 cm^3. As it evaporates, its volume decreases.

(a) It is given that the volume V cm^3 of the mothball remaining after time t weeks satisfies the differential equation
$$\frac{dV}{dt} = -\frac{1}{10}V$$
Solve this differential equation to obtain an expression for the volume remaining after t weeks.

What does this model predict for:
 (i) the volume remaining after 1 week
 (ii) the lifetime of the mothball?

(b) In an improved model, it is assumed that the radius of the mothball decreases at a constant rate. Given that, after 1 week, the volume of the mothball is 4.5 cm^3, show that the lifetime of the mothball predicted by this improved model is approximately 29 weeks.

For this model, show that the rate of loss of volume is proportional to the mothball's surface area.

[For a sphere of radius r, surface area = $4\pi r^2$, volume = $\frac{4}{3}\pi r^3$.]

SUMMARY

You should now:

- be familiar with the variables separable type of differential equation
- know how to find the general solution
- know how to deduce the particular solution given some additional information
- know how to rearrange solutions involving lns and fractions
- be able to form a differential equation from information about a rate of change
- be familiar with the terms exponential growth and decay, what these imply and the general shape of the corresponding graphs
- know how to relate connected rates of change
- be able to set up a differential equation where one factor is increasing and another decreasing

ANSWERS

Practice questions A

1. (a) $y = Ae^{x^2}$ (b) $y = \dfrac{A}{x}$
   (c) $y = Ax$ (d) $y = \pm\sqrt[4]{x^2 + A}$
   (e) $y = Ae^{-2\cos x}$ (f) $y = Ae^{-3x}$

2. (a) $y^2 = x^2 - 3$ (b) $y = \dfrac{2}{x^2 - 8}$
   (c) $y = 2e^{x^3}$ (d) $2y^2 = 2 + x - x^2$
   (e) $y = xe^{1-x^2}$

Practice questions B

1. (a) $y = \dfrac{x-1}{2-x}$ (b) $y = \dfrac{e^x}{1+e^x}$
   (c) $r = a\cos\theta$ (d) $y = \dfrac{e^{-\frac{1}{2}x^2}}{x}$
   (e) $y = \dfrac{2(x-1)}{x}e^{\frac{1}{x}}$ (f) $y = \dfrac{1-3x}{x-1}$

Practice questions C

1. (a) $\dfrac{dA}{dt} = 2\pi r \dfrac{dr}{dt}$ (b) $\dfrac{dV}{dt} = 100 \dfrac{dh}{dt}$
   (c) $\dfrac{dV}{dt} = 4\pi r^2 \dfrac{dr}{dt}$ (d) $\dfrac{dS}{dt} = 8\pi r \dfrac{dr}{dt}$

2. $\dfrac{dV}{dt} = \dfrac{1}{4}\sqrt{\left(\dfrac{S}{\pi}\right)} \dfrac{dS}{dt}$

SECTION 7

Vectors

INTRODUCTION

Using vectors, both the magnitude and direction of a quantity can be represented in a single expression. This leads to a simplification and reduction in working, especially for topics in applied mathematics and some geometrical problems in pure mathematics. In this section we shall start with vectors in two dimensions and extend our work to lines in three dimensions, the angles between them and their points of intersection, if any.

Scalars and vectors

OCR P3 5.3.7

A scientist doing some research into the effects of walking long distances might be quite interested in someone who had just completed a walk of 45 km, and even more interested if the walk was 85 km long. We can extract the most important piece of information for our purposes by asking the single question – 'how far?' This type of quantity, where only the amount or size is relevant, is called a *scalar*. A scalar can generally be represented as a point on the number line – the length of the two walks, for example, could be shown by

Figure 7.1

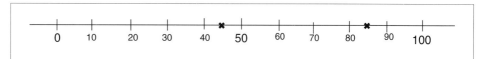

Someone studying geography might possibly be interested in the answer to more than one question – not only 'how far', but also 'and in what direction?' In this case, a single number line would not be enough – we'd have to add another dimension to give some indication of this additional piece of information. Taking our example again, if the walks were 45 km in a NE (north east) direction and 85 km in a W (west) direction, both relative to the same starting point, we could represent them thus:

Figure 7.2

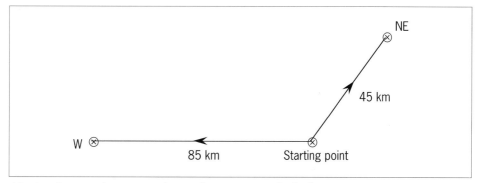

Notice that we have to indicate the sense in which the displacement took place by putting an arrow pointing the appropriate way; this is to distinguish 85 km W from 85 km E.

Quantities that answer two questions, how much and in what direction, are called *vectors*. If it's necessary to make it clear whether a particular quantity is scalar or vector, we write the scalar as a normal letter, e.g. 'a' when writing by hand or italicised '*a*', as in this book, whereas the corresponding vector is in bold type, '**a**' (printed) or underlined '<u>a</u>' (handwritten).

Now that you are clear about scalars and vectors let's see how to combine two or more vectors by adding or subtracting. Before we do this, there is one more distinction to make – between displacement vectors and position vectors.

An example of a displacement vector could be 'a distance of 10 km in a NW direction'. If N is at the top of this page and we choose the scale correctly, any of the vectors in Fig. 7.3 would fit this description, and in that sense they are all equal.

Figure 7.3

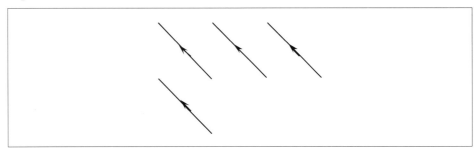

A fixed vector, or position vector, is defined relative to a given point (say *O*) and for any combination of length and direction, as above for example, there is only one vector fitting this description (see Fig. 7.4):

Figure 7.4

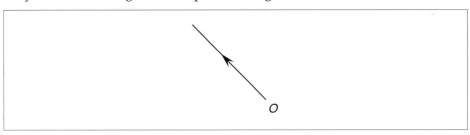

Practice questions A

1. State whether the following are scalar or vector quantities
   (a) a temperature of 50°C
   (b) a wind of 60 km/h from the north west
   (c) a height of 1.60 metres
   (d) an acceleration of 10 m/s^2
   (e) a current of 15 knots in an easterly direction
   (f) a high tide of 3.6 metres.

Adding displacement vectors OCR P3 5.3.7 (c)

Suppose we had the two displacement vectors **a** and **b**, represented by two lines, as shown in Fig.7.5:

Figure 7.5

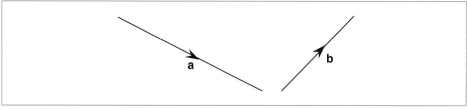

If we want to add these, we put the beginning point of **b** in contact with the end point of **a**.

Figure 7.6

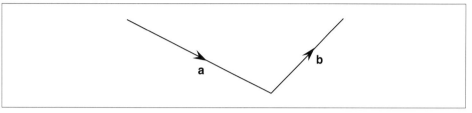

Then the sum of these is defined as the vector joining the beginning point to the end point of the 'journey'.

Figure 7.7

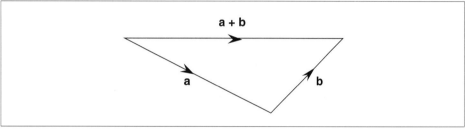

It doesn't matter how many vectors make up the 'journey', provided the tail of each vector only joins at the beginning point of another vector, so that a path can be made from the tail of the beginning vector to the head of the end vector passing through all the arrows in the same direction.

Figure 7.8

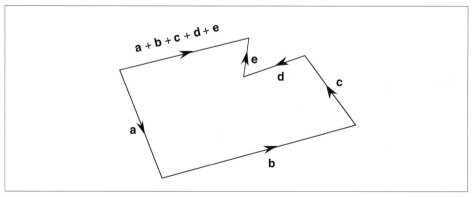

Vectors are frequently given in the form of a displacement between two points – $\overrightarrow{AB}$ is the directed line from A to B.

Figure 7.9

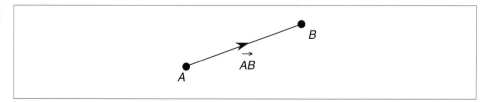

To add the vectors in this form, we have to make sure that the letter ending one vector begins another – we can add $\vec{AB}$ and $\vec{BC}$, for example:

Figure 7.10

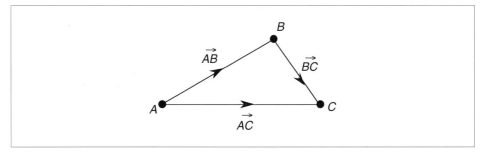

and the result will be $\vec{AC}$ i.e.

$$\vec{AB} + \vec{BC} = \vec{AC}$$

But we can't add $\vec{OP}$ and $\vec{OQ}$ (Fig. 7.11) in the same way since the letter O begins both vectors.

Figure 7.11

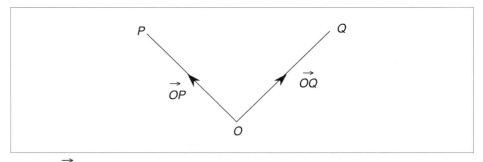

To find $\vec{PQ}$ in this case, we would have to put the letter O between the two other points:

$$\vec{PQ} = \vec{PO} + \vec{OQ}$$

and $\vec{PO}$ is $\vec{OP}$ the wrong way round, or $-\vec{OP}$:

Figure 7.12

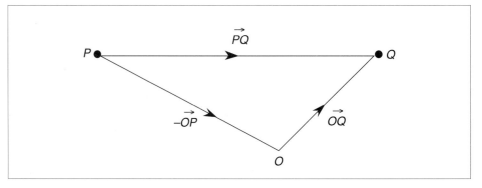

When vectors are given in relation to a fixed origin O, you would quite often write something like $\vec{OA}$ as **a** and $\vec{OS}$ as **s**, so that:

$$\vec{OP} = \mathbf{p} \quad \text{and} \quad \vec{OQ} = \mathbf{q}$$

and $\vec{PQ} = \vec{PO} + \vec{OQ}$
$= -\vec{OP} + \vec{OQ}$
$= -\mathbf{p} + \mathbf{q}$

i.e.

$$\vec{PQ} = \mathbf{q} - \mathbf{p}$$

Multiplication by a scalar

OCR P3 5.3.7 (b)

If a vector is multiplied by a *scalar* quantity, leading to $2\mathbf{a}$ or $\frac{1}{2}\mathbf{b}$, the result is a vector in the same direction but with magnitude changed according to the factor.

Figure 7.13

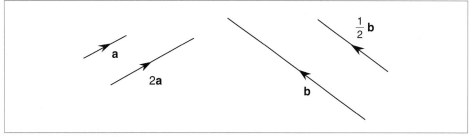

We will illustrate these points in the following example.

Example

The three non-collinear points A, B and C have position vectors $\mathbf{a}$, $\mathbf{b}$ and $\mathbf{c}$ respectively relative to an origin O. M is the mid point of BC. Find, in terms of $\mathbf{a}$, $\mathbf{b}$ and $\mathbf{c}$ as necessary,

(a) $\vec{OM}$ (b) $\vec{AM}$

Solution

A diagram is usually helpful with problems like this:

Figure 7.14

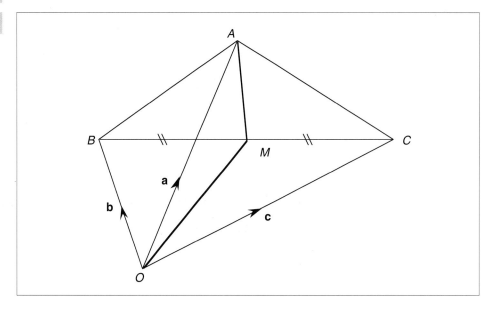

(a) We have to find a route from O to M using vectors we know. One route we could try is to go from O to B and then to M. The first half OB we know to be $\mathbf{b}$, but we don't as yet know BM. We do know that BM is half of BC and we can find BC quite easily:

$$\overrightarrow{BC} = \overrightarrow{BO} + \overrightarrow{OC}$$
$$= -\mathbf{b} + \mathbf{c}$$

so $\overrightarrow{BM} = \tfrac{1}{2}\overrightarrow{BC} = \tfrac{1}{2}(\mathbf{c} - \mathbf{b})$

and then $\overrightarrow{OM} = \overrightarrow{OB} + \overrightarrow{BM}$
$$= \mathbf{b} + \tfrac{1}{2}(\mathbf{c} - \mathbf{b})$$
$$= \mathbf{b} + \tfrac{1}{2}\mathbf{c} - \tfrac{1}{2}\mathbf{b}$$
$$= \tfrac{1}{2}\mathbf{b} + \tfrac{1}{2}\mathbf{c}$$
$$= \tfrac{1}{2}(\mathbf{b} + \mathbf{c})$$

(b) Now we can find AM without too much difficulty:

$$\overrightarrow{AM} = \overrightarrow{AO} + \overrightarrow{OM}$$
$$= -\overrightarrow{OA} + \overrightarrow{OM}$$
$$= -\mathbf{a} + \tfrac{1}{2}(\mathbf{b} + \mathbf{c}) = \tfrac{1}{2}(\mathbf{b} + \mathbf{c} - 2\mathbf{a})$$

Practice questions B

1. P has position vector $\mathbf{p}$ relative to an origin O and Q is a point such that $OQ = 2OP$. R has position vector $\mathbf{r}$ relative to the same origin and S is a point such that $OS = 2OR$. T is the mid-point of QS. Find, in terms of $\mathbf{p}$ and $\mathbf{r}$.

  (a) $\overrightarrow{PR}$ (b) $\overrightarrow{QT}$ (c) $\overrightarrow{OT}$ (d) $\overrightarrow{TR}$

The vector equation of a line

OCR P3 5.3.7 (f)

In Fig. 7.15 we have a normal (cartesian) framework divided into a grid of unit squares upon which we put a counter P. P can only change position according to the rule that it moves up twice as many spaces as it moves to the right. So if it started at the point (2, 3) and moved to the right, it would pass through the points (3, 5), (4, 7), (5, 9), etc.

Figure 7.15

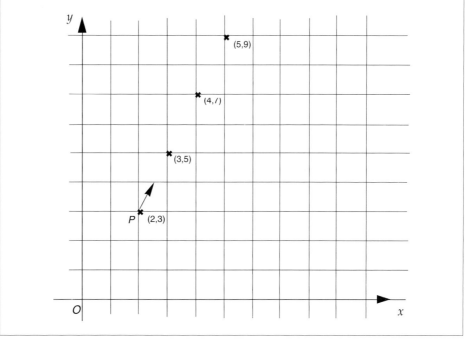

In general, after it had moved s spaces to the right it would have moved $2s$ spaces up – and its co-ordinates would be

$x = 2 + s$

$y = 3 + 2s$

These are the *parametric equations* of this particular line. Writing them in vector form combines these two equations into one equation:

$$\begin{pmatrix} x \\ y \end{pmatrix} = \begin{pmatrix} 2+s \\ 3+2s \end{pmatrix} = \begin{pmatrix} 2 \\ 3 \end{pmatrix} + s \begin{pmatrix} 1 \\ 2 \end{pmatrix}$$

Instead of writing $\begin{pmatrix} x \\ y \end{pmatrix}$ for a general position vector on the line, we usually write **r**. Our *vector equation* of this line is then

Line l_1: $\mathbf{r} = \begin{pmatrix} 2 \\ 3 \end{pmatrix} + s \begin{pmatrix} 1 \\ 2 \end{pmatrix}$ ... ①

where s is the parameter.

Conventionally, we let **i** be a unit vector in the x-direction and **j** a unit vector in the y-direction, so that we could write the equation as $\mathbf{r} = (2\mathbf{i} + 3\mathbf{j}) + s(\mathbf{i} + 2\mathbf{j})$.

The general structure of the equation:

$$\mathbf{r} \quad = \quad \mathbf{a} \quad + \quad s\,\mathbf{b}$$

General point   Position   Direction

corresponds to the familiar cartesian equation:

$$y = mx + C$$

Direction   Position

Another line could have the equation

$$l_2; \mathbf{r} = \begin{pmatrix} 4 \\ -3 \end{pmatrix} + t \begin{pmatrix} 2 \\ -1 \end{pmatrix} \text{ or } \mathbf{r} = (4\mathbf{i} - 3\mathbf{j}) + t(2\mathbf{i} - \mathbf{j})$$

and from this we can see that it passes through the point (4, –3) and for every 2 across it moves 1 down.

To find a few of the points Q passes through, we could give some different values to t;

$$t = 0, \quad \begin{pmatrix} x \\ y \end{pmatrix} = \begin{pmatrix} 4 \\ -3 \end{pmatrix} + 0 \begin{pmatrix} 2 \\ -1 \end{pmatrix} = \begin{pmatrix} 4 \\ -3 \end{pmatrix} \quad \therefore \text{ point is } (4, -3)$$

$$t = 3, \quad \begin{pmatrix} x \\ y \end{pmatrix} = \begin{pmatrix} 4 \\ -3 \end{pmatrix} + 3 \begin{pmatrix} 2 \\ -1 \end{pmatrix} = \begin{pmatrix} 10 \\ -6 \end{pmatrix} \quad \therefore \text{ point is } (10, -6)$$

$$t = -2, \quad \begin{pmatrix} x \\ y \end{pmatrix} = \begin{pmatrix} 4 \\ -3 \end{pmatrix} - 2 \begin{pmatrix} 2 \\ -1 \end{pmatrix} = \begin{pmatrix} 0 \\ -1 \end{pmatrix} \quad \therefore \text{ point is } (0, -1)$$

On the other hand, to establish whether a particular point lies on the line, we equate the x-coordinates to find the corresponding value of the parameter. We then put this value back into the equation to find the y coordinate: if this corresponds to the given point we have shown that it lies on the line.

Example Determine whether the points: (a) (3, 2) (b) (–1, –1), lie on the line with vector equation

$$\mathbf{r} = \begin{pmatrix} 2 \\ 1 \end{pmatrix} + \lambda \begin{pmatrix} 1 \\ 1 \end{pmatrix}$$

Solution
(a) Equating x-coordinates, $3 = 2 + \lambda \Rightarrow \lambda = 1$. Putting this into the line equation gives $y = 1 + \lambda = 2$ as required. (3, 2) lies on the line.

(b) For the x, $-1 = 2 + \lambda \Rightarrow \lambda = -3$

This gives $y = 1 + \lambda = 1 - 3 = -2$, not the same. (–1, –1) is not on the line.

Coordinates and position vectors

The *coordinates* of a general point in two dimensions are always written (x, y) with a comma separating the coordinates. However, the *position vector* of this general point is written in column form $\begin{pmatrix} x \\ y \end{pmatrix}$ or on one line as $x\mathbf{i} + y\mathbf{j}$.

Practice questions C

1 Find the vector equation of the line passing through:
(a) (0, 1)   with direction $(2\mathbf{i} - 5\mathbf{j})$
(b) (1, –1)   direction $(-\mathbf{i} + 2\mathbf{j})$
(c) (0, 0)   direction $(\mathbf{i} + \mathbf{j})$

2 Determine whether the following points lie on the given lines:
(a) (3, 2)   on $\mathbf{r} = 2\mathbf{i} + 6\mathbf{j} + s(\mathbf{i} - 4\mathbf{j})$
(b) (0, 0)   on $\mathbf{r} = 8\mathbf{i} + 2\mathbf{j} + s(4\mathbf{i} + \mathbf{j})$
(c) (3, –1)   on $\mathbf{r} = 5\mathbf{i} + 4\mathbf{j} + s(\mathbf{i} + 3\mathbf{j})$
(d) (–1, –3)   on $\mathbf{r} = 3\mathbf{i} + 5\mathbf{j} + s(\mathbf{i} + 2\mathbf{j})$

Equation of a line passing through two given points `OCR P3 5.3.7 (f)`

The vector equation has the two components, position and direction. If we are given the position vector of two points we can take either for the position: to find the direction of the line, we subtract the two.

Example Find the vector equation of the line passing through the points with position vectors $\mathbf{a} = 3\mathbf{i} + 2\mathbf{j}$, $\mathbf{b} = 5\mathbf{i} + 7\mathbf{j}$.

Solution See Fig. 7.16.

Figure 7.16

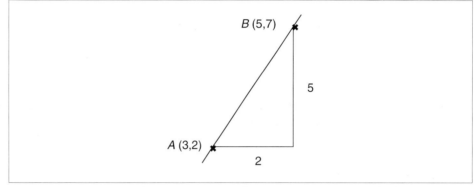

To get from A to B we have to go 5 up while moving across 2, which means that the direction vector is $\binom{2}{5}$ and the vector equation of the line passing through A and B is

$$\mathbf{r} = \binom{3}{2} + \lambda \binom{2}{5}$$

Note that this is the equation for the whole of the line passing through A and B. If we are interested only in the precise section of the line between A and B, we have to put restrictions on the set of values that the parameter λ can take. If we put $\lambda = 0$,

$$\binom{x}{y} = \binom{3}{2} + 0\binom{2}{5} = \binom{3}{2}, \text{ the position vector } \overrightarrow{OA}$$

If we put $\lambda = 1$

$$\binom{x}{y} = \binom{3}{2} + 1\binom{2}{5} = \binom{5}{7}, \text{ the position vector } \overrightarrow{OB}$$

These are the two extreme points – values of λ between 0 and 1 will give points between A and B. So if we are asked for the vector equation of the line segment AB (as opposed to the line passing through A and B), we would put

$$\mathbf{r} = \binom{3}{2} + \lambda \binom{2}{5}; \; 0 \leq \lambda \leq 1$$

In general, given the position vectors of the points, $\mathbf{a}$ and $\mathbf{b}$ for example, the direction ratios of the line joining these points is the difference between the vectors, $\mathbf{b} - \mathbf{a}$

So our equation of the line passing through the points A and B with position vectors $\mathbf{a}$ and $\mathbf{b}$ could be:

$$\mathbf{r} = \mathbf{a} + s(\mathbf{b} - \mathbf{a})$$

Note that the vector in the brackets is a *ratio*, so we can change it by multiplying or dividing through. $(6\mathbf{i} - 3\mathbf{j})$ could be $(2\mathbf{i} - \mathbf{j})$, for example, and $(-2\mathbf{i} - \mathbf{j})$ could be $(2\mathbf{i} + \mathbf{j})$.

Practice questions D

1 Find a vector equation of the line passing through the points:
 (a) (0, 1) and (4, 5) (b) (2, 3) and (−1, 2)
 (c) (−6, 6) and (2, 4) (d) (0, 0) and (2, 4)

2 Find a vector equation of the line passing through the points with position vectors:
 (a) $2\mathbf{i} + \mathbf{j}$ and $3\mathbf{i} - \mathbf{j}$ (b) $\mathbf{j}$ and $-\mathbf{i} + 4\mathbf{j}$
 (c) $-\mathbf{i} - \mathbf{j}$ and $2\mathbf{i} + 3\mathbf{j}$ (d) $3\mathbf{i}$ and $4\mathbf{j}$

Vectors in three dimensions

OCR P3 5.3.7

We've been looking at vectors in two dimensions so far – there are no complications when we add a further dimension. The direction ratios now involve three figures: the three dimensions are represented by unit vectors $\mathbf{i}$, $\mathbf{j}$ and $\mathbf{k}$. The general position vector $\mathbf{r}$ now means

$$x\mathbf{i} + y\mathbf{j} + z\mathbf{k} \quad \text{or} \quad \begin{pmatrix} x \\ y \\ z \end{pmatrix} \quad \text{in column form}$$

and the coordinates of this general point are (x, y, z).

The equation of a line in three dimensions has the same form as before, so a line passing through the points A with position vector $\mathbf{a}$ and B with position $\mathbf{b}$ is given by

$$\mathbf{r} = \mathbf{a} + \lambda(\mathbf{b} - \mathbf{a})$$

position vector   parameter   direction ratios

The line

$$\mathbf{r} = \begin{pmatrix} 1 \\ 2 \\ 3 \end{pmatrix} + \lambda \begin{pmatrix} 2 \\ -1 \\ 6 \end{pmatrix} \qquad \textit{Vector equation}$$

for example, passes through the point with coordinates, $x = 1$, $y = 2$, $z = 3$ with direction ratios $x : y : z = 2 : -1 : 6$.

Since $\mathbf{r} = \begin{pmatrix} x \\ y \\ z \end{pmatrix}$ we can separate out the three parts of this equation and put them into parametric form:

$x = 1 + 2\lambda$ *Parametric equations*
$y = 2 - \lambda$
$z = 3 + 6\lambda$

There is no single cartesian equation for this line corresponding to something like $y = 2x - 4$ in two dimensions. Instead, we have to solve these parametric equations for λ:

$$\lambda = \frac{x-1}{2} = \frac{y-2}{-1} = \frac{z-3}{6} \quad \text{Cartesian equations}$$

You have to be able to change from vector to cartesian form or the other way round and it helps at the beginning to put each of the cartesian equations equal to λ, find the parametric equations and then the vector equation. You may note that the direction ratios of the vector equation are the bottom line of the cartesian equations (i.e. $2 : -1 : 6$) and the position vector has as components the values of x, y and z which make the tops of the fractions zero, in this case 1, 2 and 3, i.e. $\begin{pmatrix} 1 \\ 2 \\ 3 \end{pmatrix}$.

Example

Put the vector equation of the line l_1

$$l_1 \; ; \; \mathbf{r} = (3\mathbf{i} - 4\mathbf{j} + 5\mathbf{k}) + \lambda (2\mathbf{i} - \mathbf{j} + \mathbf{k})$$

into   (a)   parametric form
          (b)   cartesian form.

Solution

(a) We can see this more easily in column form:

$$\begin{pmatrix} x \\ y \\ z \end{pmatrix} = \begin{pmatrix} 3 \\ -4 \\ 5 \end{pmatrix} + \lambda \begin{pmatrix} 2 \\ -1 \\ 1 \end{pmatrix} \Rightarrow \begin{array}{l} x = 3 + 2\lambda \\ y = -4 - \lambda \\ z = 5 + \lambda \end{array}$$

(b) Solving these parametric equations for λ

$$(\lambda =) \; \frac{x-3}{2} = \frac{y+4}{-1} = \frac{z-5}{1}$$

Example

Put the cartesian equations of the line l_2

$$l_2 \; : \; \frac{x+4}{2} = \frac{y-7}{3} = \frac{z+3}{-5}$$

into   (a)   parametric form
          (b)   vector form.

Solution

(a) Equating each to λ gives
$$\begin{array}{l} x = -4 + 2\lambda \\ y = 7 + 3\lambda \\ z = -3 - 5\lambda \end{array}$$

(b) In vector form, by reading down
$$\mathbf{r} = (-4\mathbf{i} + 7\mathbf{j} - 3\mathbf{k}) + \lambda (2\mathbf{i} + 3\mathbf{j} - 5\mathbf{k})$$

Special cases

There are a few particular things to note:

(a) Absence of constants: means zero(s) in the position part of the line

e.g. $\dfrac{x}{3} = \dfrac{y-2}{-1} = \dfrac{z}{4}$

is the same as $\mathbf{r} = \begin{pmatrix} 0 \\ 2 \\ 0 \end{pmatrix} + \lambda \begin{pmatrix} 3 \\ -1 \\ 4 \end{pmatrix}$

(b) Negative coefficient of x, y or z: just reverse all the signs (i.e. multiply top and bottom by -1)

e.g. $\dfrac{2-x}{4} = \dfrac{y+1}{3} = \dfrac{3-z}{2}$ becomes

$\dfrac{x-2}{-4} = \dfrac{y+1}{3} = \dfrac{z-3}{-2}$ and we can rewrite it

$\mathbf{r} = (2\mathbf{i} - \mathbf{j} + 3\mathbf{k}) + \lambda(-4\mathbf{i} + 3\mathbf{j} - 2\mathbf{k})$

(c) One of x, y or z constant: zero in the direction ratios

e.g. $\dfrac{x-4}{5} = \dfrac{z+2}{-1}$, $y = 3$

is $\mathbf{r} = (4\mathbf{i} + 3\mathbf{j} - 2\mathbf{k}) + \lambda(5\mathbf{i} - \mathbf{k})$

(d) Absence of fractions: rearrange until coefficients of x, y and z are 1 (i.e. divide top and bottom by coefficient)

e.g. $x - 3 = 2y + 5 = 3z + 2$

can be rewritten as $\dfrac{x-3}{1} = \dfrac{y + \frac{5}{2}}{\frac{1}{2}} = \dfrac{z + \frac{2}{3}}{\frac{1}{3}}$

i.e. $\mathbf{r} = \left(3\mathbf{i} - \dfrac{5}{2}\mathbf{j} - \dfrac{2}{3}\mathbf{k}\right) + \lambda\left(\mathbf{i} + \dfrac{1}{2}\mathbf{j} + \dfrac{1}{3}\mathbf{k}\right)$

The fractions in the direction part of the equation could be eliminated by multiplying through by 6 to give $\lambda(6\mathbf{i} + 3\mathbf{j} + 2\mathbf{k})$: you can't do this with the position part.

(e) Two of x, y and z constant: direction ratios have two zeros and one 1

e.g. $x = 3$, $y = 4$ $\Rightarrow \mathbf{r} = \begin{pmatrix} 3 \\ 4 \\ 0 \end{pmatrix} + \lambda \begin{pmatrix} 0 \\ 0 \\ 1 \end{pmatrix}$

The z coordinate can be any number by taking the corresponding value for λ.

Practice questions E

1 Change the following vector equations into cartesian form:

(a) $\mathbf{r} = 3\mathbf{i} + \mathbf{j} - 4\mathbf{k} + \lambda(5\mathbf{i} - 2\mathbf{j} + 3\mathbf{k})$

(b) $\mathbf{r} = 3\mathbf{i} + 2\mathbf{k} + \lambda(\mathbf{i} + \mathbf{j} - \mathbf{k})$

(c) $\mathbf{r} = \mathbf{i} + \mathbf{j} + \mathbf{k} + \lambda(\mathbf{i} - 2\mathbf{j} + \mathbf{k})$

(d) $\mathbf{r} = 2\mathbf{i} - 2\mathbf{j} + \mathbf{k} + \lambda(3\mathbf{i} - 4\mathbf{j})$

(e) $\mathbf{r} = 2\mathbf{i} + \mathbf{j} - 3\mathbf{k} + \lambda(\mathbf{i})$

(f) $\mathbf{r} = (\mathbf{i} - \mathbf{k}) + \lambda(\mathbf{i} - \mathbf{j} - \mathbf{k})$

(g) $\mathbf{r} = \lambda(\mathbf{j} + \mathbf{k})$

(h) $\mathbf{r} = \mathbf{j} - 2\mathbf{k} + \lambda \mathbf{i}$

2 Change the following into parametric form:
(a) $\mathbf{r} = 2\mathbf{i} + 3\mathbf{j} - \mathbf{k} + \lambda(5\mathbf{i} + 6\mathbf{j} - 4\mathbf{k})$
(b) $\mathbf{r} = 3\mathbf{i} + \mathbf{j} + \lambda(\mathbf{i} - 2\mathbf{j} + 2\mathbf{k})$
(c) $\mathbf{r} = \lambda(\mathbf{i} + \mathbf{j} - \mathbf{k})$
(d) $\mathbf{r} = 2\mathbf{i} + \mathbf{j} + \lambda(\mathbf{i} - \mathbf{k})$

3 Change the following into parametric form:
(a) $\dfrac{x-1}{2} = \dfrac{y+3}{5} = \dfrac{z-2}{-1}$
(b) $\dfrac{x}{5} = \dfrac{y-4}{4} = \dfrac{z+3}{-2}$
(c) $\dfrac{x-1}{4} = \dfrac{y}{-4},\ z = 2$
(d) $x = 3,\ \dfrac{y}{2} = \dfrac{z}{-1}$

4 Change the following cartesian equations into vector form:
(a) $\dfrac{x-4}{2} = \dfrac{y+3}{-1} = \dfrac{z+5}{-4}$
(b) $\dfrac{x+3}{-3} = \dfrac{1-y}{-1} = \dfrac{z}{5}$
(c) $\dfrac{x}{-1} = \dfrac{y-4}{5} = \dfrac{z}{3}$
(d) $x = 2,\ \dfrac{y-4}{7} = \dfrac{z+1}{-2}$
(e) $x = 4y = 3z$
(f) $x = 2y - 4 = 2z$
(g) $x = 3,\ y = 4$
(h) $z = 5y,\ x = 4$

5 Determine whether the given points lie on the given lines:
(a) $(3, 2, -1)$ on $\mathbf{r} = \mathbf{i} - 3\mathbf{k} + \lambda(\mathbf{i} + \mathbf{j} + \mathbf{k})$
(b) $(4, 0, -5)$ on $\mathbf{r} = -2\mathbf{j} + \mathbf{k} + \lambda(2\mathbf{i} + \mathbf{j} - 3\mathbf{k})$
(c) $(4, 1, -3)$ on $\mathbf{r} = 10\mathbf{i} + 3\mathbf{j} + \mathbf{k} + \lambda(3\mathbf{i} + \mathbf{j} - \mathbf{k})$
(d) $(5, 3, 2)$ on $\dfrac{x-1}{2} = \dfrac{y-5}{-1} = \dfrac{z+4}{3}$
(e) $(2, -1, 0)$ on $\dfrac{x+4}{-6} = y = \dfrac{z+2}{-2}$
(f) $(4, 6, -1)$ on $\dfrac{x+2}{2} = \dfrac{y-3}{1} = \dfrac{z-2}{-1}$

Intersection of two lines

OCR P3 5.3.7 (g),(h)

In two dimensions lines will always cross one another unless they are parallel. We find this point of intersection by equating the *x*-coordinates and the *y*-coordinates and solving the resulting equations in the two parameters simultaneously.

Example Find the point of intersection of the lines l_1 and l_2 with equations

$l_1:\quad \mathbf{r} = 5\mathbf{i} + 2\mathbf{j} + s(\mathbf{i} - \mathbf{j})$
$l_2:\quad \mathbf{r} = -\mathbf{i} + 2\mathbf{j} + t(4\mathbf{i} + 2\mathbf{j})$

Solution *x*-coordinates in parametric form are:

$5 + s = -1 + 4t$ ... ①

y-coords: $2 - s = 2 + 2t$ ... ②

Adding these equations, $7 = 1 + 6t \Rightarrow t = 1$ and $s = -2$

Either of these in to the appropriate equation gives the point of intersection $3\mathbf{i} + 4\mathbf{j}$.

In three dimensions lines do not necessarily meet. If you can imagine looking at two laser beams starting from different points at night, you will see that they will always appear to cross (in two dimensions) unless they are parallel. Only by looking from underneath will you be able to determine whether or not they actually meet: in other words, you have to look at the third dimension.

Exactly the same process is carried out in the abstract: a crossing point in two dimensions is found and then the corresponding coordinates in the third dimension are calculated. If these are the same the lines meet and the point can be found by substituting the value of the parameter into the relevant equation: if the coordinates are not the same, the lines do not intersect each other and are said to be *skew*.

Example

Show that the lines l_1, l_2 with equations:

l_1: $\mathbf{r} = 6\mathbf{i} - 2\mathbf{j} + 3\mathbf{k} + s(\mathbf{i} - 2\mathbf{j} + 2\mathbf{k})$
l_2: $\mathbf{r} = 3\mathbf{i} + \mathbf{j} + \mathbf{k} + t(\mathbf{i} + \mathbf{j} - 2\mathbf{k})$

intersect and determine the position vector of the point of intersection.

Solution

We find first of all a crossing point for the x- and y-coordinates:

x-coord: $6 + s = 3 + t$ ... ①
y-coord: $-2 - 2s = 1 + t$ ... ②

Subtracting, $8 + 3s = 2 \Rightarrow s = -2, t = 1$

This gives the z-coordinates l_1: $3 + 2s = 3 - 4 = -1$
l_2: $1 - 2t = 1 - 2 = -1$

These are the same, so the lines intersect. Substituting $t = 1$ into l_2, the point of intersection is $4\mathbf{i} + 2\mathbf{j} - \mathbf{k}$.

Practice questions F

1 Find the position vector of the point of intersection of the following pairs of lines

(a) $\mathbf{r} = (3\mathbf{i} + \mathbf{j}) + s(4\mathbf{i} - \mathbf{j})$ and
$\mathbf{r} = (\mathbf{i} - 2\mathbf{j}) + t(-\mathbf{i} + 2\mathbf{j})$

(b) $\mathbf{r} = (2\mathbf{i} + \mathbf{j}) + s(\mathbf{i} + 3\mathbf{j})$ and
$\mathbf{r} = (6\mathbf{i} - \mathbf{j}) + t(\mathbf{i} - 4\mathbf{j})$

(c) $\mathbf{r} = (2\mathbf{i} + 5\mathbf{j}) + s(2\mathbf{i} - 3\mathbf{j})$ and
$\mathbf{r} = (3\mathbf{i} - 3\mathbf{j}) + t(3\mathbf{i} + 2\mathbf{j})$

2 Determine whether the following pairs of lines intersect and if they do, find the position vector of their point of intersection.

(a) $\mathbf{r} = (3\mathbf{i} + 2\mathbf{j} + 3\mathbf{k}) + s(\mathbf{i} + 2\mathbf{k})$ and
$\mathbf{r} = (10\mathbf{i} - \mathbf{j} + 2\mathbf{k}) + t(3\mathbf{i} - \mathbf{j} + \mathbf{k})$

(b) $\mathbf{r} = (2\mathbf{i} + 2\mathbf{j} - 4\mathbf{k}) + s(\mathbf{i} + 3\mathbf{j} - 3\mathbf{k})$ and
$\mathbf{r} = (\mathbf{i} + \mathbf{j} + \mathbf{k}) + t(\mathbf{i} + 2\mathbf{j} - 4\mathbf{k})$

(c) $\mathbf{r} = (2\mathbf{i} + 3\mathbf{j} + \mathbf{k}) + s(2\mathbf{i} - \mathbf{j} - \mathbf{k})$ and
$\mathbf{r} = (3\mathbf{i} + 2\mathbf{k}) + t(\mathbf{i} - \mathbf{j})$

(d) $\mathbf{r} = (\mathbf{i} - \mathbf{j} + 4\mathbf{k}) + \lambda(\mathbf{i} + \mathbf{j} - \mathbf{k})$ and
$\mathbf{r} = (5\mathbf{i} + 3\mathbf{j} + \mathbf{k}) + \mu(2\mathbf{i} + 2\mathbf{j} - \mathbf{k})$

(e) $\mathbf{r} = (2\mathbf{i} + 3\mathbf{j} + 5\mathbf{k}) + \lambda(-2\mathbf{i} - \mathbf{j} - \mathbf{k})$ and
$\mathbf{r} = (-5\mathbf{i} + \mathbf{j} + \mathbf{k}) + \mu(\mathbf{i} - \mathbf{j} + \mathbf{k})$

3 Show that the following pairs of lines intersect and find their point of intersection

(a) $\dfrac{x-1}{2} = \dfrac{y-2}{3} = \dfrac{z-3}{4}$ and
$\dfrac{x-2}{1} = \dfrac{y-3}{2} = \dfrac{z-4}{3}$

(b) $\dfrac{x-2}{3} = \dfrac{y+1}{2} = \dfrac{z-1}{5}$ and
$\dfrac{x-3}{2} = 2 - y = z - 5$

(c) $\dfrac{x-3}{1} = \dfrac{y-5}{4} = \dfrac{z-2}{1}$ and
$\dfrac{x}{1} = \dfrac{y-3}{-1} = \dfrac{z-5}{-2}$

(d) $\dfrac{x-2}{-2} = \dfrac{y-3}{4} = \dfrac{z+2}{1}$ and
$\dfrac{x+6}{5} = \dfrac{y+3}{1} = \dfrac{z-1}{-2}$

(e) $\dfrac{x+2}{2} = \dfrac{y-5}{-1} = \dfrac{z+3}{2}$ and
$\dfrac{x-1}{1} = \dfrac{y-1}{2} = \dfrac{z+2}{3}$

4 Find the point of intersection of the line through the points with position vectors $(4\mathbf{i} - 2\mathbf{j} + 5\mathbf{k})$ and $(-2\mathbf{i} + \mathbf{j} - \mathbf{k})$ and the line through the points with position vectors $(-\mathbf{i} - 3\mathbf{j} - 3\mathbf{k})$ and $(8\mathbf{i} + 3\mathbf{j} + 15\mathbf{k})$.

5 Find the equation of the line passing through $(2\mathbf{i} + \mathbf{j} - 2\mathbf{k})$ and $(\mathbf{k})$ and the equation of the line passing through $(3\mathbf{i} + 2\mathbf{j} - 4\mathbf{k})$ and $(2\mathbf{i} + 4\mathbf{j} - 5\mathbf{k})$. Find the position vector of their point of intersection.

6 If the lines:
$$\mathbf{r} = (\mathbf{i} + 2\mathbf{j} + \mathbf{k}) + s(\mathbf{i} - 2\mathbf{j} + \mathbf{k})$$ and
$$\mathbf{r} = 2\mathbf{i} + t(\mathbf{i} + p\mathbf{j} + 2\mathbf{k})$$

intersect, find p and the position vector of their point of intersection.

The length of vectors

OCR P3 5.3.7 (d)

We can find the distance between the two points in two dimensions quite easily – if the points are $A(x_A, y_A)$ and $B(x_B, y_B)$, then the distance is given by Pythagoras's theorem:

Figure 7.17

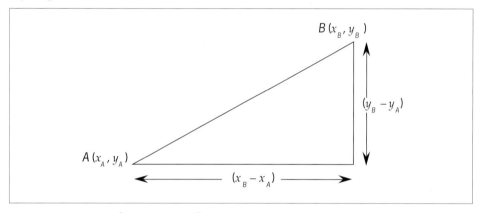

i.e. the length of $\overrightarrow{AB}$ written $|\overrightarrow{AB}|$ is given by:

$$|\overrightarrow{AB}| = \sqrt{(x_B - x_A)^2 + (y_B - y_A)^2}$$

There is quite a simple extension of this to three dimensions: if the points are $A(x_A, y_A, z_A)$ and $B(x_B, y_B, z_B)$, then the length of $\overrightarrow{AB}$ is given by:

$$|\overrightarrow{AB}| = \sqrt{(x_B - x_A)^2 + (y_B - y_A)^2 + (z_B - z_A)^2}$$

For example, the distance between the points $A(3,1,2)$ and $B(-1,2,-4)$ would be:

$$\sqrt{(-1-3)^2 + (2-1)^2 + (-4-2)^2} = \sqrt{16 + 1 + 36} = \sqrt{53}$$

Exactly the same formula applies if we are given the points A and B in the form of position vectors, i.e.

$$\overrightarrow{OA} = x_A\mathbf{i} + y_A\mathbf{j} + z_A\mathbf{k} \text{ and } \overrightarrow{OB} = x_B\mathbf{i} + y_B\mathbf{j} + z_B\mathbf{k}$$

> The length of the vector $\overrightarrow{AB}$, where A and B have position vectors
> $\begin{pmatrix} x_A \\ y_A \\ z_A \end{pmatrix}$ and $\begin{pmatrix} x_B \\ y_B \\ z_B \end{pmatrix}$ respectively, is given by
> $$|\overrightarrow{AB}| = \sqrt{(x_B - x_A)^2 + (y_B - y_A)^2 + (z_B - z_A)^2}$$

If one of these points is the origin, this formula becomes quite simple.

$$\text{If } \vec{OB} = \begin{pmatrix} x_B \\ y_B \\ z_B \end{pmatrix}, \text{ the length of } OB \text{ is given by}$$

$$|\vec{OB}| = \sqrt{x_B^2 + y_B^2 + z_B^2}$$

Here are some examples of this.

Example

Find the lengths of the following vectors.

(a) $\vec{AB}$, where $\vec{OA} = \begin{pmatrix} 1 \\ 2 \\ 3 \end{pmatrix}$ and $\vec{OB} = \begin{pmatrix} 3 \\ 5 \\ 7 \end{pmatrix}$

(b) $\vec{PQ}$, where $\vec{OP} = -\mathbf{i} + 2\mathbf{j} + 2\mathbf{k}$ and $\vec{OQ} = -3\mathbf{j} + \mathbf{j} + 4\mathbf{k}$

(c) $\vec{OR}$, where $\vec{OR} = \begin{pmatrix} -1 \\ 5 \\ 2 \end{pmatrix}$

Solution

(a) $\vec{AB} = \vec{OB} - \vec{OA} = \begin{pmatrix} 2 \\ 3 \\ 4 \end{pmatrix}$. By the formula, this has length

$\sqrt{2^2 + 3^2 + 4^2} = \sqrt{29}$ units.

(b) $\vec{PQ} = \vec{OQ} - \vec{OP} = -2\mathbf{i} - \mathbf{j} + 2\mathbf{k}$. This has length

$|\vec{PQ}| = \sqrt{(-2)^2 + (-1)^2 + 2^2} = 3$ units

(c) From the formula

$|\vec{OR}| = \sqrt{(-1)^2 + 5^2 + 2^2} = \sqrt{30}$ units

Unit vectors

OCR P3 5.3.7 (c)

A unit vector has a length (or magnitude) of one. If we are asked to find a unit vector parallel to a given vector, we simply find the magnitude of the given vector and divide by this magnitude: the result is the unit vector.

Example

Find a unit vector parallel to $\mathbf{a} = \mathbf{i} - 2\mathbf{j} + 2\mathbf{k}$.

Solution

The magnitude of $|\mathbf{a}|$ is $\sqrt{1^2 + (-2)^2 + 2^2} = 3$

$\Rightarrow$ a unit vector parallel to $\mathbf{a}$ (or in the direction of $\mathbf{a}$) is

$\frac{1}{3}(\mathbf{i} - 2\mathbf{j} + 2\mathbf{k})$

Symbolically, this is written $\hat{\mathbf{a}}$, i.e. $\hat{\mathbf{a}} = \dfrac{\mathbf{a}}{|\mathbf{a}|}$.

The same applies when we want to find a vector of any magnitude parallel to a given vector: we find the magnitude of the given vector and adjust accordingly.

Example Find a vector of magnitude 21 in the direction of **b** = –2**i** + 3**j** + 6**k**

Solution The magnitude of **b** is $\sqrt{(-2)^2 + 3^2 + 6^2} = 7$

| **b** | = 7 and we want a magnitude of 21,

i.e. 3 | **b** | ⇒ the vector is 3 (–2**i** + 3**j** + 6**k**)

i.e. –6**i** + 9**j** + 18**k**

Practice questions G

1 Find the magnitude of the following vectors:
(a) 2**i** – **j** – 2**k** (b) 6**i** + 2**j** – 3**k**
(c) 3**i** + 4**k** (d) **i** – **j** + **k**
(e) 4**i** + 4**j** – 2**k** (f) 3**k**

2 Find unit vectors parallel to the following:
(a) 2**i** – 6**j** + 3**k** (b) 6**i** – 3**j** + 6**k**
(c) 4**i** – 3**j** (d) 5**j**

3 Find vectors of the given magnitude, parallel to the given vector:
(a) magnitude 10, parallel to 4**j** + 3**k**
(b) 18, to **i** – 2**j** + 2**k**
(c) 3.5, to 2**i** + 6**j** + 3**k**
(d) $\sqrt{8}$, to **i** + **k**

We're now going to see if we can find the angle between any two vectors, but before we do this, we need to look at a way of multiplying vectors, which is called the *scalar product*, because the result is a scalar quantity.

The scalar product

OCR P3 5.3.7 (e)

Suppose we have two vectors **a** and **b**, and a light is shining directly down onto **b**. The length of the shadow of **a** on **b** is called the *projection* of **a** on **b**. This projection is the distance that vector **a** moves in the direction **b**.

Figure 7.18

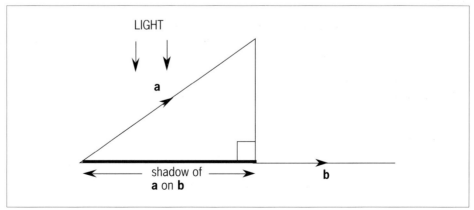

Then the scalar product of **a** and **b**, which can also be called the *dot product* because it is written **a** . **b**, is defined by:

a . **b** = projection of **a** on **b** × length of **b**   … [*]

In Fig. 7.19 we have marked the lengths of the vectors **a** and **b** and the angle between them, which we can call θ.

Figure 7.19

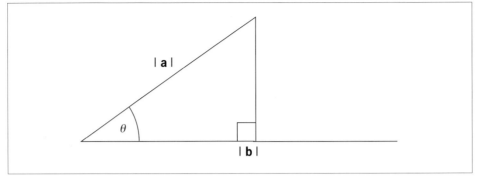

Since the triangle is right-angled, we have:

$$\cos\theta = \frac{\text{adjacent}}{\text{hypotenuse}} = \frac{\text{projection of }\mathbf{a}\text{ on }\mathbf{b}}{|\mathbf{a}|}$$

Rearranging, projection of **a** on **b** = $|\mathbf{a}|\cos\theta$

and putting this into [*] gives:

$$\mathbf{a}.\mathbf{b} = |\mathbf{a}|\cos\theta \times |\mathbf{b}|$$
$$= |\mathbf{a}||\mathbf{b}|\cos\theta \qquad \ldots \text{①}$$

This is one way of defining the scalar product but there is another way, direct from the components of the two vectors. Suppose the two vectors **a** and **b** were $\begin{pmatrix} x_a \\ y_a \\ z_a \end{pmatrix}$ and $\begin{pmatrix} x_b \\ y_b \\ z_b \end{pmatrix}$ respectively

Then $\mathbf{a}.\mathbf{b} = x_a x_b + y_a y_b + z_a z_b \qquad \ldots \text{②}$

For example, if **a** is $\begin{pmatrix} 2 \\ 1 \\ 2 \end{pmatrix}$ and **b** is $\begin{pmatrix} 3 \\ -2 \\ 8 \end{pmatrix}$ then

$$\mathbf{a}.\mathbf{b} = \begin{pmatrix} 2 \\ 1 \\ 2 \end{pmatrix}.\begin{pmatrix} 3 \\ -2 \\ 8 \end{pmatrix} = 2 \times 3 + 1 \times -2 + 2 \times 8 = 6 - 2 + 16 = 20$$

Note that we end up with a *scalar* (i.e. a number): hence the name scalar product.

The angle between vectors

OCR P3 5.3.7 (e)

Combining these two results ① and ②, we can find the angle θ between the vectors **a** and **b**.

> The angle θ between two vectors **a** and **b**, $\begin{pmatrix} x_a \\ y_a \\ z_a \end{pmatrix}$ and $\begin{pmatrix} x_b \\ y_b \\ z_b \end{pmatrix}$ respectively,
>
> is given by $\cos\theta = \dfrac{\mathbf{a}.\mathbf{b}}{|\mathbf{a}||\mathbf{b}|}$
>
> where $\mathbf{a}.\mathbf{b} = x_a x_b + y_a y_b + z_a z_b$

Example

Find the angle between the vectors:

(a) $\mathbf{a} = \begin{pmatrix} 3 \\ -1 \\ 2 \end{pmatrix}$ and $\mathbf{b} = \begin{pmatrix} 2 \\ 1 \\ -3 \end{pmatrix}$

(b) $\mathbf{p} = 3\mathbf{i} + 2\mathbf{j} - 6\mathbf{k}$ and $\mathbf{q} = -\mathbf{j} + \mathbf{k}$

(c) $\mathbf{u} = \begin{pmatrix} a \\ -2a \\ -5a \end{pmatrix}$ and $\mathbf{v} = \begin{pmatrix} -2a \\ 7a \\ 5a \end{pmatrix}$, where a is a scalar

Solution

(a) The angle is given by:

$$\cos\theta = \frac{\mathbf{a} \cdot \mathbf{b}}{|\mathbf{a}||\mathbf{b}|} = \frac{(3 \times 2) + (-1 \times 1) + (2 \times -3)}{\sqrt{3^2 + (-1)^2 + 2^2} \sqrt{2^2 + 1^2 + (-3)^2}}$$

$$= \frac{6 - 1 - 6}{\sqrt{14} \sqrt{14}}$$

$$= \frac{-1}{14}$$

so the angle θ is 94.1° (correct to one decimal place)

(b) Here, $\cos\theta = \dfrac{3 \times 0 + 2 \times -1 + -6 \times 1}{\sqrt{3^2 + 2^2 + (-6)^2} \sqrt{(-1)^2 + 1^2}} = \dfrac{-8}{7\sqrt{2}}$

and the angle is 143.9° (to one decimal place)

(c) $\mathbf{u} \cdot \mathbf{v} = -2a^2 - 14a^2 - 25a^2 \quad = -41a^2$

$|\mathbf{u}| = \sqrt{a^2 + 4a^2 + 25a^2} \quad = \sqrt{30a^2} \quad = a\sqrt{30}$

$|\mathbf{v}| = \sqrt{4a^2 + 49a^2 + 25a^2} \quad = \sqrt{78a^2} \quad = a\sqrt{78}$

and $\cos\theta = \dfrac{-41a^2}{a\sqrt{30}\, a\sqrt{78}} = \dfrac{-41}{\sqrt{30}\, \sqrt{78}}$

The angle between these vectors is 147.9° (to one decimal place).

Note: the dot product was negative in these examples, giving an angle of more than 90°. This is possible because the angle between two vectors takes into account the direction of the vectors – both vectors have to be pointed away from the vertex.

Figure 7.20

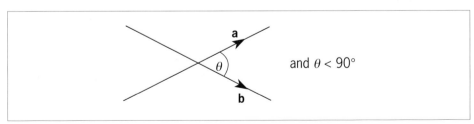

and $\theta < 90°$

Figure 7.21

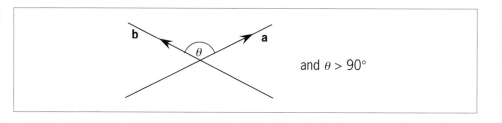

and $\theta > 90°$

Angles between lines

OCR P3 5.3.7 (h)

To find the angle between two lines, we take only the *direction ratios* part of the equation. Since the lines do not have a direction, there are two possible (complementary) angles: conventionally we take the acute angle (θ in the diagram)

Figure 7.22

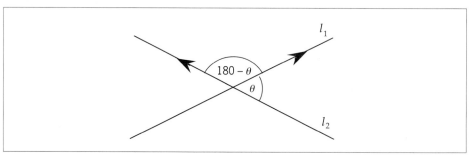

Example Find the angle between

l_1: $\mathbf{r} = (2\mathbf{i} + \mathbf{j} - 3\mathbf{k}) + \lambda(2\mathbf{i} + 3\mathbf{j} - \mathbf{k})$

l_2: $\mathbf{r} = (5\mathbf{i} + \mathbf{j}) + \mu(4\mathbf{i} - \mathbf{j} - 2\mathbf{k})$

Solution

$$\cos\theta = \frac{(2\mathbf{i} + 3\mathbf{j} - \mathbf{k}).(4\mathbf{i} - \mathbf{j} - 2\mathbf{k})}{\sqrt{2^2 + 3^2 + (-1)^2} \times \sqrt{4^2 + (-1)^2 + (-2)^2}} = \frac{8 - 3 + 2}{\sqrt{14}\sqrt{21}}$$

$$= \frac{1}{\sqrt{6}} \Rightarrow \theta = 65.9° \text{ (1 dp)}$$

Practice questions H

1 Find the value of the following scalar products:

(a) $(3\mathbf{i} - 2\mathbf{j} + 3\mathbf{k}).(4\mathbf{i} + 2\mathbf{j} - \mathbf{k})$

(b) $(3\mathbf{i} + \mathbf{j}).(\mathbf{i} + \mathbf{j} + \mathbf{k})$

(c) $(5\mathbf{i} + 2\mathbf{j} - 3\mathbf{k}).(2\mathbf{i} + \mathbf{j} - 7\mathbf{k})$

(d) $(\mathbf{i} + \mathbf{j}).(\mathbf{i} - \mathbf{j})$

(e) $(2\mathbf{i} - 3\mathbf{j} + 6\mathbf{k}).(\mathbf{j} - 3\mathbf{k})$

(f) $(\mathbf{i} + 4\mathbf{k}).(\mathbf{j} - \mathbf{k})$

2 Find the angles between the following pairs of vectors, giving your answer in degrees to 1 d.p., where appropriate:

(a) $(2\mathbf{i} + 3\mathbf{j} - \mathbf{k})$ and $(\mathbf{i} + \mathbf{j} - 4\mathbf{k})$

(b) $(\mathbf{i} + 2\mathbf{j} - 2\mathbf{k})$ and $(2\mathbf{i} - 6\mathbf{j} + 3\mathbf{k})$

(c) $(3\mathbf{i} + 4\mathbf{j})$ and $(\mathbf{i} - 7\mathbf{k})$

(d) $(\mathbf{i} + \mathbf{j} + \mathbf{k})$ and $(\mathbf{i} + \mathbf{j} - \mathbf{k})$

(e) $(3\mathbf{i} + 2\mathbf{j} - \mathbf{k})$ and $(2\mathbf{i} - \mathbf{j} + 4\mathbf{k})$

3 Find the acute angle between the following pairs of lines, giving your answers in degrees to 1 d.p. where appropriate:

(a) $\mathbf{r} = (3\mathbf{i} + 2\mathbf{j} + 2\mathbf{k}) + s\,(\mathbf{i} + 2\mathbf{k})$ and
$\mathbf{r} = (10\mathbf{i} - \mathbf{j} + 2\mathbf{k}) + t\,(3\mathbf{i} - \mathbf{j} + \mathbf{k})$

(b) $\mathbf{r} = (2\mathbf{i} + 2\mathbf{j} - 4\mathbf{k}) + s\,(\mathbf{i} + 3\mathbf{j} - 3\mathbf{k})$ and
$\mathbf{r} = (\mathbf{i} + \mathbf{j} + \mathbf{k}) + t\,(\mathbf{i} + 2\mathbf{j} - 4\mathbf{k})$

(c) $\mathbf{r} = (2\mathbf{i} + 3\mathbf{j} + \mathbf{k}) + s\,(2\mathbf{i} - \mathbf{j} - \mathbf{k})$ and
$\mathbf{r} = (3\mathbf{i} + 2\mathbf{k}) + t\,(\mathbf{i} - \mathbf{j})$

(d) $\mathbf{r} = (\mathbf{i} - \mathbf{j} + 4\mathbf{k}) + \lambda\,(\mathbf{i} + \mathbf{j} - \mathbf{k})$ and
$\mathbf{r} = (5\mathbf{i} + 3\mathbf{j} + \mathbf{k}) + \mu\,(2\mathbf{i} + 2\mathbf{j} - \mathbf{k})$

(e) $\mathbf{r} = (2\mathbf{i} + 3\mathbf{j} + 5\mathbf{k}) + \lambda\,(2\mathbf{i} - \mathbf{j} - \mathbf{k})$ and
$\mathbf{r} = (-5\mathbf{i} + \mathbf{j} + \mathbf{k}) + \mu\,(\mathbf{i} - \mathbf{j} + \mathbf{k})$

4 Find the acute angle between the following pairs of lines, giving your answers in degrees to 1 d.p.

(a) $\dfrac{x-1}{2} = \dfrac{y-2}{3} = \dfrac{z-3}{4}$ and
$\dfrac{x-2}{1} = \dfrac{y-3}{2} = \dfrac{z-4}{3}$

(b) $\dfrac{x-2}{3} = \dfrac{y+1}{2} = \dfrac{z-1}{5}$ and
$\dfrac{x-3}{2} = 2-y = z-5$

(c) $\dfrac{x-3}{1} = \dfrac{y-5}{4} = \dfrac{z-2}{1}$ and
$\dfrac{x}{1} = \dfrac{y-3}{-1} = \dfrac{z-5}{-2}$

(d) $\dfrac{x-2}{-2} = \dfrac{y-3}{4} = \dfrac{z+2}{1}$ and
$\dfrac{x+6}{5} = \dfrac{y+3}{1} = \dfrac{z-1}{-2}$

(e) $\dfrac{x+2}{2} = \dfrac{y-5}{-1} = \dfrac{z+3}{2}$ and
$\dfrac{x-1}{1} = \dfrac{y-1}{2} = \dfrac{z+2}{3}$

5 With respect to a fixed origin O, the lines l_1 and l_2 are given by the equations:

$l_1: \mathbf{r} = (2\mathbf{i} + 3\mathbf{j} - 2\mathbf{k}) + \lambda(-2\mathbf{i} + 4\mathbf{j} + \mathbf{k})$
$l_2: \mathbf{r} = (-6\mathbf{i} - 3\mathbf{j} + \mathbf{k}) + \mu(5\mathbf{i} + \mathbf{j} - 2\mathbf{k})$

where λ and μ are scalar parameters.

(a) Show that l_1 and l_2 meet and find the position vector of their point of intersection.

(b) Find, to the nearest $0.1°$, the acute angle between l_1 and l_2.

6 The points A and B have position vectors $\mathbf{i} + 3\mathbf{j} + 4\mathbf{k}$ and $2\mathbf{i} + \mathbf{j} + 6\mathbf{k}$ respectively, relative to an origin O.

(a) Find a vector equation of the line AB.

The line CD has vector equation

$\mathbf{r} = (2\mathbf{i} + 2\mathbf{j} + 2\mathbf{k}) + \mu(\mathbf{i} + 2\mathbf{j} + 2\mathbf{k})$

where μ is a real parameter.

(b) Show that the lines AB and CD do not intersect.

(c) Find the acute angle between these two lines, giving your answer to the nearest degree.

7 The position vectors of three points A, B and C with respect to a fixed origin O are $2\mathbf{i} - 2\mathbf{j} + \mathbf{k}$, $4\mathbf{i} + 2\mathbf{j} + \mathbf{k}$ and $\mathbf{i} + \mathbf{j} + 3\mathbf{k}$ respectively.

Find unit vectors in the directions of $\overrightarrow{CA}$ and $\overrightarrow{CB}$. Calculate angle ACB in degrees, correct to 1 decimal place.

Perpendicular vectors

OCR P3 5.3.7 (e)

The formula by which we calculate the angle makes it easy to see when two vectors are perpendicular to each other. In this case the angle between them is, of course, $90°$.

$$\cos 90° = \frac{\mathbf{a} \cdot \mathbf{b}}{|\mathbf{a}||\mathbf{b}|} \quad \text{but } \cos 90° = 0$$

i.e. $\mathbf{a} \cdot \mathbf{b} = 0$

This is a very important result which we will emphasise by putting it in a box.

If $\mathbf{a}$ and $\mathbf{b}$ are two non-zero vectors

$\mathbf{a} \cdot \mathbf{b} = 0 \Leftrightarrow \mathbf{a}$ and $\mathbf{b}$ are mutually perpendicular

---

The double-ended implication sign, ⇔, means that the two statements amount to exactly the same thing – if either one is true, so is the other.

Example

Find the value of p such that the vectors $\mathbf{a} = 3\mathbf{i} + \mathbf{j} - 2\mathbf{k}$ and $\mathbf{b} = \mathbf{i} + p\mathbf{j} + 4\mathbf{k}$ are perpendicular.

Solution

We need $\mathbf{a} \cdot \mathbf{b} = 0$

$$\Rightarrow (3\mathbf{i} + \mathbf{j} - 2\mathbf{k}) \cdot (\mathbf{i} + p\mathbf{j} + 4\mathbf{k}) = 0$$
$$3 + p - 8 = 0 \Rightarrow p = 5$$

Example

Find the value of a such that the lines

$l_1: \mathbf{r} = (2\mathbf{i} + 4\mathbf{j} - \mathbf{k}) + \lambda(5\mathbf{i} + \mathbf{j} + a\mathbf{k})$
$l_2: \mathbf{r} = (3\mathbf{i} - \mathbf{j}) + \mu(2\mathbf{i} - 4\mathbf{j} + 3\mathbf{k})$

are perpendicular.

Solution

The direction ratios need to be perpendicular

$$(5\mathbf{i} + \mathbf{j} + a\mathbf{k}) \cdot (2\mathbf{i} - 4\mathbf{j} + 3\mathbf{k}) = 0$$
$$10 - 4 + 3a = 0$$
$$\Rightarrow a = -2$$

Perpendicular from a point onto a line

We can use this property to find the foot of a perpendicular from a point onto a line.

Figure 7.23

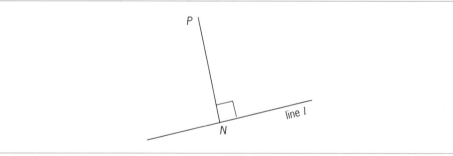

The procedure is:

(a) find the general position vector of N

(b) find the vector $\overrightarrow{PN}$ if necessary

(c) solve $\overrightarrow{PN} \cdot \text{line} = 0$

(d) substitute this value into (a).

Example

Find the position vector of the foot N of the perpendicular from the origin O onto the line with equation

$$\mathbf{r} = (\mathbf{i} + 3\mathbf{j} + 4\mathbf{k}) + \lambda(3\mathbf{i} + \mathbf{j} + 2\mathbf{k})$$

Solution

(a) Since N is on the line, its position vector will be $(\mathbf{i} + 3\mathbf{j} + 4\mathbf{k}) + n(3\mathbf{i} + \mathbf{j} + 2\mathbf{k})$ for some particular value n of the parameter λ, i.e.

$$\vec{ON} = (1 + 3n)\mathbf{i} + (3 + n)\mathbf{j} + (4 + 2n)\mathbf{k}$$

(b) The point P is the origin: we already have $\vec{ON}$

(c) If $\vec{ON}$ is perpendicular to the line,

$$[(1 + 3n)\mathbf{i} + (3 + n)\mathbf{j} + (4 + 2n)\mathbf{k}] \cdot [3\mathbf{i} + \mathbf{j} + 2\mathbf{k}] = 0$$

i.e.
$$3 + 9n + 3 + n + 8 + 4n = 0$$
$$14 + 14n = 0$$
$$\Rightarrow n = -1$$

(d) Put this into (a): $\vec{ON} = -2\mathbf{i} + 2\mathbf{j} + 2\mathbf{k}$

Here's another example where P is not the origin.

Example

Find the position vector of the foot N of the perpendicular from the point P with position vector $-3\mathbf{i} + \mathbf{j}$ onto the line with equation

$$\mathbf{r} = (3\mathbf{i} + 5\mathbf{j} + 5\mathbf{k}) + \lambda(2\mathbf{i} - \mathbf{j} + 2\mathbf{k})$$

Find also the length of the perpendicular.

Solution

(a) N is on the line

$$\Rightarrow \vec{ON} = (3 + 2n)\mathbf{i} + (5 - n)\mathbf{j} + (5 + 2n)\mathbf{k}$$

(b) $\vec{PN} = \vec{ON} - \vec{OP} = (6 + 2n)\mathbf{i} + (4 - n)\mathbf{j} + (5 + 2n)\mathbf{k}$

(c) $\vec{PN} \cdot \text{line} = 0$

$$\Rightarrow [(6 + 2n)\mathbf{i} + (4 - n)\mathbf{j} + (5 + 2n)\mathbf{k}] \cdot [2\mathbf{i} - \mathbf{j} + 2\mathbf{k}] = 0$$

$$12 + 4n - 4 + n + 10 + 4n = 0$$

$$18 + 9n = 0 \Rightarrow n = -2$$

(d) $\vec{ON} = -\mathbf{i} + 7\mathbf{j} + \mathbf{k}$

The length of the perpendicular is $|\vec{PN}|$, so put the $n = -2$ into $\vec{PN}$ in (b)

$$\vec{PN} = 2\mathbf{i} + 6\mathbf{j} + \mathbf{k} \Rightarrow |\vec{PN}| = \sqrt{4 + 36 + 1} = \sqrt{41}$$

Practice questions I

1. Find the value(s) of p for which the following pairs of vectors are perpendicular
   (a) $2\mathbf{i} + 3\mathbf{j} + 4\mathbf{k}$ and $p\mathbf{i} - 2\mathbf{j} + 4\mathbf{k}$
   (b) $p\mathbf{i} + \mathbf{j} - \mathbf{k}$ and $\mathbf{i} + 2\mathbf{j} + 4\mathbf{k}$
   (c) $-3\mathbf{i} + \mathbf{j} + p\mathbf{k}$ and $2\mathbf{i} - p\mathbf{j} + p\mathbf{k}$
   (d) $-2\mathbf{i} + 5\mathbf{j} - p\mathbf{k}$ and $-2\mathbf{i} + p\mathbf{k}$

Section 7

2 Find the value of p for which the following pairs of lines are perpendicular:

(a) $\mathbf{r} = \mathbf{i} - 4\mathbf{j} + \mathbf{k} + \lambda(\mathbf{i} - 4\mathbf{j} + 8\mathbf{k})$ and
$\mathbf{r} = (3\mathbf{i} - 2\mathbf{j} + \mathbf{k}) + \mu(-4\mathbf{i} + 9\mathbf{j} + p\mathbf{k})$

(b) $\mathbf{r} = 3\mathbf{i} + 4\mathbf{j} + \lambda(3\mathbf{i} + p\mathbf{j} + 6\mathbf{k})$ and
$\mathbf{r} = (2\mathbf{j} - 5\mathbf{k}) + \mu(-4\mathbf{i} + 9\mathbf{j} + 5\mathbf{k})$

(c) $\mathbf{r} = 5\mathbf{i} + \lambda(2\mathbf{i} + p\mathbf{j} + \mathbf{k})$ and
$\mathbf{r} = (-3\mathbf{i} + 2\mathbf{j} + \mathbf{k}) + \mu(3\mathbf{i} - 2\mathbf{j} + 6\mathbf{k})$

(d) $\mathbf{r} = 5\mathbf{i} - 2\mathbf{j} + 4\mathbf{k} + \lambda(3\mathbf{i} + \mathbf{j} + \mathbf{k})$ and
$\mathbf{r} = (\mathbf{i} - 5\mathbf{k}) + \mu(2\mathbf{i} + 4\mathbf{j} + p\mathbf{k})$

3 Show that the vector $\mathbf{i} + \mathbf{j} + \mathbf{k}$ is perpendicular to both the lines:

$\mathbf{r} = (2\mathbf{i} + 3\mathbf{j}) + \lambda(2\mathbf{i} - \mathbf{j} - \mathbf{k})$ and
$\mathbf{r} = (3\mathbf{i} + 4\mathbf{j} - \mathbf{k}) + \mu(3\mathbf{i} + \mathbf{j} - 4\mathbf{k})$

4 Find the position vector of the point P on the following lines such that OP is perpendicular to the line:

(a) $\mathbf{r} = (-\mathbf{i} - 4\mathbf{j} + 7\mathbf{k}) + \lambda(\mathbf{i} - 2\mathbf{j} + 3\mathbf{k})$

(b) $\mathbf{r} = (4\mathbf{i} - 7\mathbf{j} + 10\mathbf{k}) + \lambda(3\mathbf{i} + 3\mathbf{j} - \mathbf{k})$

(c) $\mathbf{r} = (5\mathbf{i} + 3\mathbf{j} - \mathbf{k}) + \lambda(-\mathbf{i} + \mathbf{j} + \mathbf{k})$

(d) $\mathbf{r} = (10\mathbf{i} - 5\mathbf{j} + \mathbf{k}) + \lambda(2\mathbf{i} - \mathbf{j} + 2\mathbf{k})$

5 Find the position vector of the point P on the following lines such that AP is perpendicular to the line:

(a) $(4\mathbf{i} + 2\mathbf{j} + 7\mathbf{k}) + \lambda(2\mathbf{i} - 3\mathbf{j} + 4\mathbf{k})$:
$\overrightarrow{OA} = -2\mathbf{i} + 12\mathbf{j} + 3\mathbf{k}$

(b) $(6\mathbf{i} + 5\mathbf{j} - \mathbf{k}) + \lambda(2\mathbf{i} + 3\mathbf{k})$:
$\overrightarrow{OA} = \mathbf{i} + 3\mathbf{j} - 2\mathbf{k}$

(c) $(4\mathbf{i} + \mathbf{j} + 3\mathbf{k}) + \lambda(2\mathbf{i} - \mathbf{j} + \mathbf{k})$: $\overrightarrow{OA} = -\mathbf{i} + 6\mathbf{j}$

(d) $(-4\mathbf{i} + 7\mathbf{j}) + \lambda(3\mathbf{i} - \mathbf{j} + \mathbf{k})$: $\overrightarrow{OA} = 4\mathbf{i} + \mathbf{j} + 3\mathbf{k}$

6 With respect to an origin O, the points A and C have position vectors $3\mathbf{i} + 11\mathbf{j} + 11\mathbf{k}$ and $2\mathbf{i} - \mathbf{j} + 10\mathbf{k}$ respectively. The point B is such that $OABC$ is a parallelogram.

(a) Write down the position vector of B.

(b) Find the size of the angle OAB, giving your answer to the nearest degree.

(c) Find a vector equation of the line AB.

(d) Find the position vector of the point on the line AB closest to O.

[AQA(AEB) 1994]

7 A is the point $(3, 2, 1)$ and B is the point $(5, 4, 0)$.

(a) Find a vector equation for the line AB

(b) Find the position vector of the point P on the line such that OP is perpendicular to the line AB

(c) Find the length of OP

8 Find a non-zero vector which is perpendicular to both the vectors $2\mathbf{i} + \mathbf{j} + \mathbf{k}$ and $\mathbf{i} - \mathbf{j} + 5\mathbf{k}$.

9 Referred to a fixed origin O, the lines l_1 and l_2 have equations:

$l_1: \mathbf{r} = (\mathbf{i} + \mathbf{j} + \mathbf{k}) + \lambda(\mathbf{i} - 7\mathbf{j} + 2\mathbf{k})$

$l_2: \mathbf{r} = (8\mathbf{i} - 4\mathbf{j} + 7\mathbf{k}) + u(3\mathbf{i} + \mathbf{j} + 2\mathbf{k})$

where λ and μ are scalar parameters.

Show that the lines l_1 and l_2 are perpendicular and that they intersect at the point P whose position vector is $2\mathbf{i} - 6\mathbf{j} + 3\mathbf{k}$.

SUMMARY EXERCISES

1 The vectors $\mathbf{u}$ and $\mathbf{v}$ are given by

$\mathbf{u} = 5\mathbf{i} - 4\mathbf{j} + s\mathbf{k}$, $\mathbf{v} = 2\mathbf{i} + t\mathbf{j} - 3\mathbf{k}$

(a) Given that the vectors $\mathbf{u}$ and $\mathbf{v}$ are perpendicular, find a relation between the scalars s and t.

(b) Given instead that the vectors $\mathbf{u}$ and $\mathbf{v}$ are parallel, find the values of the scalars s and t.

2 The line l passes through the points with position vectors $\mathbf{i} + 2\mathbf{j} + 3\mathbf{k}$ and $\mathbf{i} + 6\mathbf{j}$ relative to an origin O.

(a) Find an equation for l in vector form.

The line m has equation

$\mathbf{r} = 3\mathbf{i} + 6\mathbf{j} + \mathbf{k} + \lambda(\mathbf{i} - 2\mathbf{j} + 2\mathbf{k})$.

(b) Find the acute angle between l and m, giving your answer to the nearest degree.

3 Find by calculation the position vector of the point of intersection of the lines

$$\mathbf{r} = \begin{pmatrix} 1 \\ 2 \end{pmatrix} + s \begin{pmatrix} 1 \\ -1 \end{pmatrix} \text{ and } \mathbf{r} = \begin{pmatrix} -2 \\ 1 \end{pmatrix} + t \begin{pmatrix} -3 \\ 1 \end{pmatrix},$$

where s and t are real parameters.

Find also the acute angle between the lines, giving your answer correct to the nearest 0.1°.

4 With respect to an origin O, the position vectors of the points L and M are $2\mathbf{i} - 3\mathbf{j} + 3\mathbf{k}$ and $5\mathbf{i} + \mathbf{j} + c\mathbf{k}$ respectively, where c is a constant.

The point N is such that $OLMN$ is a rectangle.

(a) Find the value of c.

(b) Write down the position vector of N.

(c) Find, in the form $\mathbf{r} = \mathbf{p} + t\mathbf{q}$, an equation of the line MN.

5 Referred to a fixed origin O, the points A and B have position vectors

$$5\mathbf{i} + \mathbf{j} + 2\mathbf{k} \text{ and } -\mathbf{i} + 7\mathbf{j} + 8\mathbf{k}$$

The line l_1 passes through A and the line l_2 passes through B. The lines l_1 and l_2 intersect at the point C whose position vector is $\mathbf{i} + 2\mathbf{j} + \mathbf{k}$

(a) Find equations for the lines l_1 and l_2, giving each in the form $\mathbf{r} = \mathbf{a} + t\mathbf{b}$

(b) Find the size of $\angle ACB$, giving your answer to the nearest degree.

6 The points P and Q have coordinates $(1, 6, 1)$ and $(4, 0, -8)$ respectively. Find an equation, in the form $\mathbf{r} = \mathbf{a} + t\mathbf{b}$, for the straight line through P and Q.

The line l, which passes through the origin, has equation

$$\mathbf{r} = s \begin{pmatrix} 1 \\ 2 \\ -1 \end{pmatrix}.$$

Write down three equations which must be satisfied by the real parameters s and t if the lines l and PQ intersect. Find the values of s and t satisfying these three equations, and hence find the coordinates of the point of intersection of the two lines.

The line m has equation

$$\mathbf{r} = \begin{pmatrix} 0 \\ 1 \\ 1 \end{pmatrix} + u \begin{pmatrix} 5 \\ a \\ 5 \end{pmatrix},$$

where u is a real parameter and a is a constant.

Find the positive value of a for which the angle between l and m is 60°.

7 The points $A(24, 6, 0)$, $B(30, 12, 12)$ and $C(18, 6, 36)$ are referred to cartesian axes, origin O.

(a) Find a vector equation for the line passing through the points A and B.

The point P lies on the line passing through A and B.

(b) Show that $\overrightarrow{CP}$ can be expressed as $(6 + t)\mathbf{i} + t\mathbf{j} + (2t - 36)\mathbf{k}$, where t is a parameter.

(c) Given that $\overrightarrow{CP}$ is perpendicular to $\overrightarrow{AB}$, find the coordinates of P.

(d) Hence, or otherwise, find the area of the triangle ABC, giving your answer to 3 significant figures.

8 The equation of a straight line l is

$$\mathbf{r} = \begin{pmatrix} 1 \\ 2 \\ 3 \end{pmatrix} + t \begin{pmatrix} -1 \\ 1 \\ 1 \end{pmatrix},$$

where t is a parameter. The point A on l is given by $t = 0$, and the origin of position vectors is O.

(a) Calculate the acute angle between OA and l, giving your answer correct to the nearest degree

(b) Find the position vector of the point P on l such that OP is perpendicular to l.

(c) A point Q on l is such that the length of OQ is 5 units. Find the two possible position vectors of Q.

(d) The points R and S on l are given by $t = \lambda$ and $t = 2\lambda$ respectively. Show that there is no value of λ for which OR and OS are perpendicular.

9 Referred to a fixed origin O, the points P, Q and R have position vectors $(2\mathbf{i} + \mathbf{j} + \mathbf{k})$, $(5\mathbf{j} + 3\mathbf{k})$ and $(5\mathbf{i} - 4\mathbf{j} + 2\mathbf{k})$ respectively.

(a) Find in the form $\mathbf{r} = \mathbf{a} + t\mathbf{b}$, an equation of the line PQ.

(b) Show that the point S with position vector $(4\mathbf{i} - 3\mathbf{j} - \mathbf{k})$ lies on PQ.

(c) Show that the lines PQ and RS are perpendicular.

(d) Find the size of $\angle PQR$, giving your answer to 0.1°.

10 The point A has coordinates $(7, -1, 3)$ and the point B has coordinates $(10, -2, 2)$. The line l has vector equation $\mathbf{r} = \mathbf{i} + \mathbf{j} + \mathbf{k} + \lambda(3\mathbf{i} - \mathbf{j} + \mathbf{k})$, where λ is a real parameter

  (a) Show that the point A lies on the line l.

  (b) Find the length of AB.

  (c) Find the size of the acute angle between the line l and the line segment AB, giving your answer to the nearest degree.

  (d) Hence, or otherwise, calculate the perpendicular distance from B to the line l, giving your answer to 2 significant figures.

11 Vectors $\mathbf{r}$ and $\mathbf{s}$ are given by

$$\mathbf{r} = \lambda\mathbf{i} + (2\lambda - 1)\mathbf{j} - \mathbf{k}$$

$$\mathbf{s} = (1 - \lambda)\mathbf{i} + 3\lambda\mathbf{j} + (4\lambda - 1)\mathbf{k}$$

where λ is a scalar.

  (a) Find the values of λ for which $\mathbf{r}$ and $\mathbf{s}$ are perpendicular.

  When $\lambda = 2$, $\mathbf{r}$ and $\mathbf{s}$ are the position vectors of the points A and B respectively, referred to an origin O.

  (b) Find $\overrightarrow{AB}$.

  (c) Use a scalar product to find the size of $\angle BAO$, giving your answer to the nearest degree.

12 Referred to a fixed origin, the line l_1 has equation $\mathbf{r} = 12\mathbf{i} + 5\mathbf{j} + \mathbf{k} + \lambda(2\mathbf{i} + \mathbf{j} + \mathbf{k})$ and the line l_2 has equation

$$\mathbf{r} = \mathbf{i} + \mathbf{j} - 2\mathbf{k} + \mu(3\mathbf{i} - \mathbf{k})$$

  (a) Show that the lines l_1 and l_2 intersect and find the position vector of A, their point of intersection.

  (b) Find, to the nearest degree, the acute angle between the lines l_1 and l_2.

  The point B with position vector $16\mathbf{i} + 7\mathbf{j} + 3\mathbf{k}$ lies on the line l_1. The point C with position vector $22\mathbf{i} + \mathbf{j} - 9\mathbf{k}$ and the point D both lies on the line l_2.

  (c) Given that BD is perpendicular to AC, find the position vector of D.

  (d) Hence, or otherwise, prove that $\triangle ABC$ is isosceles.

13 (a) Find a vector equation of the straight line l through the points with position vectors $2\mathbf{i} + \mathbf{j} + 4\mathbf{k}$ and $3\mathbf{i} + 2\mathbf{k}$, relative to the origin O.

  The line m has cartesian equations

  $$7 - x = \frac{y + 6}{2} = z + 4$$

  (b) Write down a vector equation of m.

  (c) Show that the lines l and m meet and find the position vector of their point of intersection.

  The distinct points P and Q lie on l and are each a distance $\sqrt{17}$ units from O.

  (d) Find the position vectors of P and Q.

14 The points A and B have position vectors $3\mathbf{i} + 2\mathbf{j} + \mathbf{k}$ and $\mathbf{i} + 2\mathbf{j} + 3\mathbf{k}$, respectively, relative to the origin O. The point C is on the line OA produced and is such that $AC = 2\,OA$. The point D is on OB produced and is such that $BD = OB$. The point X is such that $OCXD$ is a parallelogram. Show that the line AX is parallel to the vector $\mathbf{i} + \mathbf{j} + \mathbf{k}$.

Find, in the form $\mathbf{r} = \mathbf{u} + t\mathbf{v}$, the equations of the lines AX and CD.

Give a reason why the lines AX and CD intersect and find the position vector of the point of intersection.

Find the angle BAX.

15 The lines l_1, l_2 and l_3 are given by

l_1: $\mathbf{r} = 10\mathbf{i} + \mathbf{j} + 9\mathbf{k} + \mu(3\mathbf{i} + \mathbf{j} + 4\mathbf{k})$

l_2: $x = \dfrac{y + 9}{2} = \dfrac{z - 13}{-3}$

l_3: $\mathbf{r} = -3\mathbf{i} - 5\mathbf{j} - 4\mathbf{k} + \lambda(4\mathbf{i} + 3\mathbf{j} + \mathbf{k})$

where μ and λ are parameters.

  (a) Show that the point $A(4, -1, 1)$ lies on both l_1 and l_2.

  (b) Rewrite the equation for l_2 in the form $\mathbf{r} = \mathbf{a} + v\mathbf{b}$, where v is a parameter.

  (c) Show that l_2 and l_3 intersect at the point $C(1, -2, -3)$.

  (d) Show that $AC = BC$.

  (e) Find the size of angle ACB, giving your answer in degrees to the nearest degree.

  (f) Write down the coordinates of the point D on AB such that CD is perpendicular to AB.

16 A line l_1 passes through the point A, with position vector $5\mathbf{i} + 3\mathbf{j}$, and the point B, with position vector $-2\mathbf{i} - 4\mathbf{j} + 7\mathbf{k}$

(a) Write down an equation of the line l_1.

A second line l_2 has equation
$\mathbf{r} = \mathbf{i} - 3\mathbf{j} - 4\mathbf{k} + \mu(\mathbf{i} + 2\mathbf{j} + 3\mathbf{k})$ where μ is a parameter.

(b) Show that l_1 and l_2 are perpendicular to each other.

(c) Show that the two lines meet, and find the position vector of the point of intersection.

The point C has position vector $2\mathbf{i} - \mathbf{j} - \mathbf{k}$.

(d) Show that C lies on l_2.

The point D is the image of C after reflection in the line l_1.

(e) Find the position vector of D.

SUMMARY

Having finished this section, you should now:

- appreciate the difference between a scalar and a vector
- be able to add and subtract vectors
- know the geometrical significance of multiplying a vector by a scalar
- know that the vector equation of a line is of the form $\mathbf{r} =$ position $+ \lambda$ (direction)
- be able to find the vector equation of a line passing through two given points
- be able to determine whether a given point lies on a given line
- be able to generalise the work in 2D to lines in 3D
- be able to change from a vector equation to cartesian equations and vice versa via the parametric equations
- be able to determine whether two lines intersect
- be able to determine the position vector of the point of intersection
- be able to find the magnitude of a vector
- be able to find unit vectors, or vectors of any given magnitude, parallel to a given vector
- be able to find the scalar product of two vectors
- be able to use this to find the angle between vectors and between lines
- know the condition for two vectors or lines to be perpendicular
- be able to find the position vector of the foot of a perpendicular from a given point onto a given line
- be feeling quite pleased that you've come to the end of this long and quite tricky section.

ANSWERS

Practice questions A

1. (a) scalar (b) vector
   (c) scalar (d) scalar
   (e) vector (f) scalar

Practice questions B

2. (a) $-\mathbf{p}+\mathbf{r}$ (b) $-\mathbf{p}+\mathbf{r}$
   (c) $\mathbf{p}+\mathbf{r}$ (d) $-\mathbf{p}$

Practice questions C

1. (a) $\mathbf{r}=\mathbf{j}+s\,(2\mathbf{i}-5\mathbf{j})$
   (b) $\mathbf{r}=\mathbf{i}-\mathbf{j}+s\,(-\mathbf{i}+2\mathbf{j})$
   (c) $\mathbf{r}=s\,(\mathbf{i}+\mathbf{j})$
2. (a) Yes (b) Yes (c) No
   (d) Yes

Practice questions D

1. (a) $\mathbf{r}=\mathbf{j}+s\,(\mathbf{i}+\mathbf{j})$
   [direction reduced from $4\mathbf{i}+4\mathbf{j}$]
   (b) $\mathbf{r}=2\mathbf{i}+3\mathbf{j}+s\,(3\mathbf{i}+\mathbf{j})$
   [direction from $-3\mathbf{i}-\mathbf{j}$]
   (c) $\mathbf{r}=-6\mathbf{i}+6\mathbf{j}+s\,(4\mathbf{i}-\mathbf{j})$
   (d) $\mathbf{r}=s\,(\mathbf{i}+2\mathbf{j})$
2. (a) $\mathbf{r}=2\mathbf{i}+\mathbf{j}+s\,(\mathbf{i}-2\mathbf{j})$
   (b) $\mathbf{r}=\mathbf{j}+s\,(\mathbf{i}-3\mathbf{j})$
   (c) $\mathbf{r}=-\mathbf{i}-\mathbf{j}+s\,(3\mathbf{i}+4\mathbf{j})$
   (d) $\mathbf{r}=3\mathbf{i}+s\,(3\mathbf{i}-4\mathbf{j})$

Practice questions E

1. (a) $\dfrac{x-3}{5}=\dfrac{y-1}{-2}=\dfrac{z+4}{3}$
   (b) $\dfrac{x-3}{1}=\dfrac{y}{1}=\dfrac{z-2}{-1}$
   (c) $\dfrac{x-1}{1}=\dfrac{y-1}{-2}=\dfrac{z-1}{1}$
   (d) $\dfrac{x-2}{3}=\dfrac{y+2}{-4},\ z=1$
   (e) $y=1,\ z=-3$
   (f) $\dfrac{x-1}{1}=\dfrac{y}{-1}=\dfrac{z+1}{-1}$
   (g) $x=0,\ y=z$
   (h) $y=1,\ z=-2$

2. (a) $x=2+5\lambda$ (b) $x=3+\lambda$
       $y=3+6\lambda$ $y=1-2\lambda$
       $z=-1-4\lambda$ $z=2\lambda$
   (c) $x=\lambda$ (d) $x=2+\lambda$
       $y=\lambda$ $y=1$
       $z=-\lambda$ $z=-\lambda$

3. (a) $x=1+2\lambda$ (b) $x=5\lambda$
       $y=-3+5\lambda$ $y=4+4\lambda$
       $z=2-\lambda$ $z=-3-2\lambda$
   (c) $x=1+4\lambda$ (d) $x=3$
       $y=-4\lambda$ $y=2\lambda$
       $z=2$ $z=-\lambda$

4. (a) $\mathbf{r}=4\mathbf{i}-3\mathbf{j}-5\mathbf{k}+\lambda\,(2\mathbf{i}-\mathbf{j}-4\mathbf{k})$
   (b) $\mathbf{r}=-3\mathbf{i}+\mathbf{j}+\lambda\,(-3\mathbf{i}+\mathbf{j}+5\mathbf{k})$
   (c) $\mathbf{r}=4\mathbf{j}+\lambda\,(-\mathbf{i}+5\mathbf{j}+3\mathbf{k})$
   (d) $\mathbf{r}=2\mathbf{i}+4\mathbf{j}-\mathbf{k}+\lambda\,(7\mathbf{j}-2\mathbf{k})$
   (e) $\mathbf{r}=\lambda\,(12\mathbf{i}+3\mathbf{j}+4\mathbf{k})$
   (f) $\mathbf{r}=2\mathbf{j}+\lambda\,(2\mathbf{i}+\mathbf{j}+\mathbf{k})$
   (g) $\mathbf{r}=3\mathbf{i}+4\mathbf{j}+\lambda\mathbf{k}$
   (h) $\mathbf{r}=4\mathbf{i}+\lambda\,(\mathbf{j}+5\mathbf{k})$

5. (a) Yes (b) Yes (c) No (d) Yes
   (e) Yes (f) Yes

Practice questions F

1. (a) $-\mathbf{i}+2\mathbf{j}$ (b) $4\mathbf{i}+7\mathbf{j}$ (c) $6\mathbf{i}-\mathbf{j}$
2. (a) $\mathbf{i}+2\mathbf{j}-\mathbf{k}$ (b) $3\mathbf{i}+5\mathbf{j}-7\mathbf{k}$
   (c) don't intersect (d) $3\mathbf{i}+\mathbf{j}+2\mathbf{k}$
   (e) $-4\mathbf{i}+2\mathbf{k}$
3. (a) $(3, 5, 7)$ (b) $(5, 1, 6)$
   (c) $(2, 1, 1)$ (d) $(4, -1, -3)$
   (e) $(2, 3, 1)$
4. $2\mathbf{i}-\mathbf{j}+3\mathbf{k}$
5. $\mathbf{r}=(2\mathbf{i}+\mathbf{j}-2\mathbf{k})+s\,(2\mathbf{i}+\mathbf{j}-3\mathbf{k})$ and
   $\mathbf{r}=(3\mathbf{i}+2\mathbf{j}-4\mathbf{k})+t\,(\mathbf{i}-2\mathbf{j}+\mathbf{k})$
   $\dfrac{16}{5}\mathbf{i}+\dfrac{8}{5}\mathbf{j}-\dfrac{19}{5}\mathbf{k}$
6. $p=-2,\ 4\mathbf{i}-4\mathbf{j}+4\mathbf{k}$

Practice questions G

1 (a) 3 (b) 7 (c) 5 (d) √3 (e) 6 (f) 3

2 (a) $\frac{1}{7}(2\mathbf{i} - 6\mathbf{j} + 3\mathbf{k})$ (b) $\frac{1}{3}(2\mathbf{i} - \mathbf{j} + 2\mathbf{k})$

   (c) $\frac{1}{5}(4\mathbf{i} - 3\mathbf{j})$ (d) $\mathbf{j}$

3 (a) $8\mathbf{j} + 6\mathbf{k}$ (b) $6\mathbf{i} - 12\mathbf{j} + 12\mathbf{k}$

   (c) $\mathbf{i} + 3\mathbf{j} + \frac{3}{2}\mathbf{k}$ (d) $2\mathbf{i} + 2\mathbf{k}$

Practice questions H

1 (a) 5 (b) 4 (c) 33
   (d) 0 (e) −21 (f) −4

2 (a) 55.5° (b) 139.6° (c) 85.1°
   (d) 70.5° (e) 90°

3 (a) 47.6° (b) 18.0° (c) 30°
   (d) 15.8° (e) 61.9°

4 (a) 7.0° (b) 53.4° (c) 61.2°
   (d) 71.4° (e) 57.7°

5 a) $4\mathbf{i} - \mathbf{j} - 3\mathbf{k}$ (b) 71.4°

6 (a) $\mathbf{r} = (\mathbf{i} + 3\mathbf{j} + 4\mathbf{k}) + \lambda(\mathbf{i} - 2\mathbf{j} + 2\mathbf{k})$

   (b) e.g. $x \alpha y \Rightarrow z = \frac{3}{2}$ and $\frac{11}{2}$

   (c) 84°

7 $\frac{1}{\sqrt{14}}(\mathbf{i} - 3\mathbf{j} - 2\mathbf{k})$, $\frac{1}{\sqrt{14}}(3\mathbf{i} + \mathbf{j} - 2\mathbf{k})$, 73.4°

Practice questions I

1 (a) −5 (b) 2 (c) 3, −2 (d) ±2

2 (a) 5 (b) −2 (c) 6 (d) −10

4 (a) $-3\mathbf{i} + \mathbf{k}$ (b) $7\mathbf{i} - 4\mathbf{j} + 9\mathbf{k}$
   (c) $4\mathbf{i} + 4\mathbf{j}$ (d) $4\mathbf{i} - 2\mathbf{j} - 5\mathbf{k}$

5 (a) $8\mathbf{j} - \mathbf{k}$ (b) $4\mathbf{i} + 5\mathbf{j} - 4\mathbf{k}$
   (c) $-2\mathbf{i} + 4\mathbf{j}$ (d) $5\mathbf{i} + 4\mathbf{j} + 3\mathbf{k}$

6 (a) $5\mathbf{i} + 10\mathbf{j} + 21\mathbf{k}$ (b) 130°
   (c) $\mathbf{r} = (3\mathbf{i} + 11\mathbf{j} + 11\mathbf{k}) + \lambda(2\mathbf{i} - \mathbf{j} + 10\mathbf{k})$
   (d) $\mathbf{i} + 12\mathbf{j} + \mathbf{k}$

7 (a) $\mathbf{r} = (3\mathbf{i} + 2\mathbf{j} + \mathbf{k}) + \lambda(2\mathbf{i} + 2\mathbf{j} - \mathbf{k})$
   (b) $\mathbf{i} - 2\mathbf{k}$ (c) √5

8 Any factor of the form $a(-2\mathbf{i} + 3\mathbf{j} + \mathbf{k})$ where a is constant.

P3
Practice examination paper

Attempt all 8 questions. Available marks are given in brackets at the end of each question.

1. Find the centre and radius of the circle
$$x^2 + 4x + y^2 - 6y = 4$$ [3]

2. When the polynomial $f(x) = x^3 + ax^2 + bx - 10$ is divided by $(x+1)$ and $(x-1)$, the remainders are -6 and -10 respectively.

   Find the values of a and b. [4]

3. Using the substitution $u = \cos x$ or otherwise, find:
$$\int \cos^4 x \sin 2x \, dx$$ [6]

4. $f(x) \equiv \dfrac{3x}{e^{2x}}$

   (a) Find $f'(x)$. [4]

   (b) Find $\displaystyle\int_0^1 f(x) \, dx$, giving your answer in exact form. [6]

5. The lines l_1 and l_2 have vector equations:

   l_1: $\mathbf{r} = (10\mathbf{i} - \mathbf{j} + 2\mathbf{k}) + \lambda\,(3\mathbf{i} - \mathbf{j} + \mathbf{k})$

   l_2: $\mathbf{r} = (3\mathbf{i} + 2\mathbf{j} + 3\mathbf{k}) + \mu\,(\mathbf{i} + 2\mathbf{k})$

   (a) Show that the lines intersect and find the point of intersection. [6]

   (b) Find the acute angle between the lines, giving your answer in degrees to 1 decimal place. [4]

6. (a) Given that $f(x) = \ln(\sec x + \tan x)$, find and simplify an expression for $f'(x)$, giving your answer as a single trigonometric ratio. [5]

   (b) Solve the differential equation
$$\frac{1}{y \cos x} \frac{dy}{dx} = 1$$
   given that $y = 2$ when $x = 0$.

   Give your answer in a form which does not contain logarithms. [7]

7. A curve C is defined by the parametric equations
$$x = t^2 - 1, \quad y = 2t + 1 \qquad t \in \mathbb{R}$$

   (a) Find an equation of the normal N to the curve at the point where $t = 2$. [3]

   (b) Sketch the curve. [5]

   (c) Find the area of the finite region bounded by C, the normal N and the positive x- and y-axes. [6]

8. (a) Express
$$f(x) \equiv \frac{1 + x - x^2}{(1 - 2x)(1 + x^2)}, \qquad |x| < \frac{1}{2}$$
   in partial fractions. [4]

   (b) Find a series expansion for $f(x)$ in ascending powers of x, up to and including the term in x^3. [6]

   (c) Find $\displaystyle\int_0^{1/4} f(x) \, dx$, giving your answer in the form $\dfrac{1}{2} \ln k$, where k is rational. [6]

P3

Solutions

Section 1

1 $f(x) \equiv ax^2 + bx + c \equiv a(x-1)(x-2) + px + q$
$f(2) = 1 \Rightarrow a(1)(0) + 2p + q = 1 \Rightarrow 2p + q = 1$
$f(1) = 2 \Rightarrow a(0)(-1) + p + q = 2 \Rightarrow p + q = 2$
From these, $p = -1, q = 3$

2 $f(-1) = 4 \Rightarrow -2 + a - b - 3 = 4, a - b = 9$
$f(3) = 84 \Rightarrow 54 + 9a + 3b - 3 = 84, \quad 9a + 3b = 33$
$\qquad\qquad\qquad\qquad\qquad\qquad\qquad 3a + b = 11$
$a = 5, b = -4$
$2x^3 + 5x^2 - 4x - 3 = (x-1)(2x^2 + 7x + 3)$
$\qquad\qquad\qquad\quad = (x-1)(2x+1)(x+3)$

3 (a) $f(k) = k \Rightarrow k^3 + k^2 - 2k + 1 = k$
$\qquad 2k^3 - 3k + 1 = 0$
$\qquad (k-1)(2k^2 + 2k - 1) = 0$
$\qquad 2k^2 + 2k - 1 = 0 \Rightarrow k = \dfrac{-2 \pm \sqrt{4+8}}{2}$
$\qquad\qquad\qquad\qquad\qquad = -1 \pm \sqrt{3}$
$\qquad \Rightarrow k = 1, -1 + \sqrt{3} \text{ or } -1 - \sqrt{3}$

(b) $p(x) \equiv (x-1)(x-2) q(x) + Ax + B$
$p(1) = 5 \Rightarrow A + B = 5$
$p(2) = 7 \Rightarrow 2A + B = 7 \Rightarrow A = 2, B = 3$
When $p(x)$ is divided by $(x-1)(x-2)$, the quotient is $q(x)$ and the remainder $2x + 3$

4 (a) $f(-2) = -24 - 8 - 2p - 6$
i.e. remainder is $-38 - 2p$
If $x + 2$ factor, $-38 - 2p = 0 \Rightarrow p = -19$

(b) $(x-2)(x^2 + 2x - 8) = 0$
$(x-2)(x+4)(x-2) = 0$
$\Rightarrow x = 2 \text{ or } x = -4$

(c) $f(1) = f(-2)$
$\Rightarrow 1 + a + b + c = -8 + 4a - 2b + c$
$\Rightarrow 9 + 3b = 3a$
$\qquad a = b + 3 \qquad\qquad\qquad \ldots \text{①}$
$f(-1) = 3 \Rightarrow -1 + a - b + c = 3$
From ①, $a - b = 3 \Rightarrow -1 + 3 + c = 3 \Rightarrow c = 1$

5 (a) $f(-3) = 10 \Rightarrow -27 + 18 - 3a + 4 = 10$
$\qquad -3a = 15, a = -5$
$f\left(\dfrac{3}{2}\right) = \dfrac{27}{8} + \dfrac{9}{2} - \dfrac{15}{2} + 4 = \dfrac{35}{8}$

(b) $2x^3 + 5x^2 + x - 2 = 0$
$(x + 1)(2x^2 + 3x - 2) = 0$

$(x + 1)(2x - 1)(x + 2) = 0$
$x = -1, \dfrac{1}{2} \text{ or } -2$

(c) $f(1) = p \Rightarrow 1 + a + b = p$
$f(2) = p + 6 \Rightarrow 4 + 2a + b = p + 6$
$\qquad\qquad\qquad \Rightarrow 3 + a = 6, a = 3$

6 (a) $f(1) = 1 - 3 + 2 - 2 + 3 + 1 = 2 \neq 0$
$\qquad \Rightarrow (x - 1) \text{ not factor}$
$f(-1) = -1 - 3 - 2 - 2 - 3 + 1 = -10 \neq 0$
$\qquad \Rightarrow (x + 1) \text{ not factor}$

(b) $f(1) = a + b = 2 \qquad$ from above
$f(-1) = -a + b = -10 \quad$ from above
$\Rightarrow 2b = -8, b = -4, a = 6$
Then remainder is $6x - 4$

(c)
$\qquad\qquad\qquad\qquad x^3 - 3x^2 + x + 1$
$x^2 + 1 \overline{\smash{\big)} x^5 - 3x^4 + 2x^3 - 2x^2 + 3x + 1}$
$\qquad\qquad\quad \underline{x^5 \qquad\quad + x^3}$
$\qquad\qquad\qquad -3x^4 + x^3 - 2x^2$
$\qquad\qquad\qquad \underline{-3x^4 \qquad\quad - 3x^2}$
$\qquad\qquad\qquad\qquad\quad x^3 + x^2 + 3x$
$\qquad\qquad\qquad\qquad\quad \underline{x^3 \qquad + x}$
$\qquad\qquad\qquad\qquad\qquad\quad x^2 + 2x + 1$
$\qquad\qquad\qquad\qquad\qquad\quad \underline{x^2 \qquad + 1}$
$\qquad\qquad\qquad\qquad\quad \text{Remainder} \quad 2x$

(d) From above,
$f(x) = (x^2 + 1)(x^3 - 3x^2 + x + 1) + 2x$
If $f(x) = 2x \Rightarrow f(x) - 2x = 0$
$\Rightarrow (x^2 + 1)(x^3 - 3x^2 + x + 1) = 0$
$x^2 + 1 \neq 0, (x - 1)(x^2 - 2x - 1) = 0$
$x^2 - 2x - 1 = 0 \Rightarrow x = \dfrac{2 \pm \sqrt{4+4}}{2} = 1 \pm \sqrt{2}$
i.e. $x = 1, 1 + \sqrt{2} \text{ or } 1 - \sqrt{2}$

7 (a) $P(1) = 0 \Rightarrow x - 1$ factor
$P(-1) = 0 \Rightarrow x + 1$ factor

(b) $Q(x) = x^4 + 4x^3 + ax^2 + bx + 5$
$\qquad\quad = (x^2 - 1) R(x) + 2x + 3$
$Q(1) = 1 + 4 + a + b + 5 = 2 + 3 \Rightarrow a + b = -5$
$Q(-1) = 1 - 4 + a - b + 5 = -2 + 3 \Rightarrow a - b = -1$
$a = -3, b = -2$

(c) Since $(x - 1)$ and $(x + 1)$ are factors of $P(x)$,
$(x - 1)(x + 1) = x^2 - 1$ is a factor.
$\Rightarrow$ remainder is just $4(2x + 3) = 8x + 12$

137

Section 2

1 (a) Completing the square, $(x-3)^2 + (y+2)^2 - 9 - 4 = 7$
$$\Rightarrow (x-3)^2 + (y+2)^2 = 20$$
Centre (3, −2), radius $\sqrt{20}$

(b) When $y = 0$, $x^2 - 6x + 7 \Rightarrow x^2 - 6x - 7 = 0$
$$(x-7)(x+1) = 0$$
$$\Rightarrow x = 7 \text{ or } x = -1$$

(c) Calling $A(7, 0)$ and $B(-1, 0)$, with $C(3, -2)$
$AB = 8$, $AC = \sqrt{4^2 + 2^2} = \sqrt{20}$, $BC = \sqrt{4^2 + 2^2} = \sqrt{20}$
$\cos A\hat{B}C = \dfrac{64 + 20 - 20}{2 \times 8 \times \sqrt{20}} = \dfrac{64}{16\sqrt{20}} = \dfrac{2}{\sqrt{5}}$

2 (a) $(x-2)^2 + (y+3)^2 - 4 - 9 = 12$
$$\Rightarrow (x-2)^2 + (y+3)^2 = 5^2 = \text{centre } (2, -3), \text{ radius } 5$$

(b) If arc $PQ = 10$, $r\theta = 10$ and with $r = 5$, $\theta = 2$
where θ is the angle POQ

Then area, $\dfrac{1}{2} r^2 \theta = \dfrac{1}{2} \times 5^2 \times 2 = 25$

3 (a) $(x+1)^2 + (y-4)^2 - 1 - 16 = 152$
$(x+1)^2 + (y-4)^2 = 13^2$: centre (−1, 4), radius 13

(b) (i) If $y = 16$, $x = k \Rightarrow k^2 + 16^2 + 2k - 8 \times 16 = 152$
$k^2 + 256 + 2k - 128 = 152$
$k^2 + 2k - 24 = 0$
$(k+6)(k-4) = 0$
$\Rightarrow k = 4$ (since > 0)

(ii)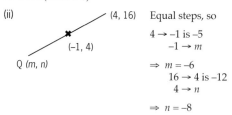

Equal steps, so
$4 \to -1$ is -5
$-1 \to m$
$\Rightarrow m = -6$
$16 \to 4$ is -12
$4 \to n$
$\Rightarrow n = -8$

i.e. coordinates of Q are $(-6, -8)$

4 $(x+2)^2 + (y-3)^2 = 5^2$: centre (−2, 3), radius 5
$y = 0 \Rightarrow x^2 + 4x - 12 = 0$
$\Rightarrow x = -6 \text{ or } x = 2$

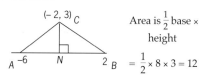

Area is $\dfrac{1}{2}$ base × height
$= \dfrac{1}{2} \times 8 \times 3 = 12$

By trigonometry, $\tan N\hat{C}A = \dfrac{4}{3} \Rightarrow A\hat{C}B = 1.8546$

$\Rightarrow$ total sector is $\dfrac{1}{2} \times 5^2 \times 1.8546 = 28.18$
Subtract $\triangle$, area is $28.18 - 12 = 16.2$ (3 s.f.)

5 $x^2 + (y-4)^2 = 5^2 \Rightarrow$ centre (0, 4), radius 5
Substituting $(3y + k)^2 + y^2 - 8y = 9$
$9y^2 + 6yk + k^2 + y^2 - 8y = 9$
$10y^2 + (6k - 8)y + (k^2 - 9) = 0$ as required
Need discriminant zero, $(6k - 8)^2 - 4 \times 10 \times (k^2 - 9) = 0$
$36k^2 - 96k + 64 - 40k^2 + 360 = 0$
$-4k^2 - 96k + 424 = 0$
$k^2 + 24k - 106 = 0$
$k = \dfrac{-24 \pm \sqrt{576 + 424}}{2} = \dfrac{-24 \pm \sqrt{1000}}{2}$
$= -12 \pm 5\sqrt{10}$

6 (a) Grad $\dfrac{4}{-2} = -2$
$\Rightarrow y - 2 = -2(x - 3)$, $2x + y = 8$

(b) (i) Substituting, $9 + 4 - 24 + 2a + b = 0$
$$\Rightarrow 2a + b = 11 \quad \ldots \text{①}$$
$1 + 36 - 8 + 6a + b = 0$
$$\Rightarrow 6a + b = -29 \quad \ldots \text{②}$$
Solving $a = -10$, $b = 31$

(ii) Centre (by completing the square) is (4, 5)
$(x - 4)^2 + (y - 5)^2 + 31 - 25 - 16 = 0$
$\Rightarrow$ radius is $\sqrt{10}$

7 Grad PQ is $\dfrac{\dfrac{c}{q} - \dfrac{c}{p}}{cq - cp} = \dfrac{\dfrac{1}{q} - \dfrac{1}{p}}{q - p}$ (÷ by c)
$= \dfrac{\dfrac{p - q}{pq}}{q - p} = \dfrac{p - q}{pq} \times \dfrac{1}{q - p} = \dfrac{-1}{pq}$

when q approaches p, we have the gradient of the tangent at p, where $q = p$, is $\dfrac{-1}{p^2}$

$\Rightarrow y - \dfrac{c}{p} = -\dfrac{1}{p^2}(x - cp) \Rightarrow p^2 y - cp = -x + cp$
i.e. $p^2 y + x = 2cp$

8 $d^2 = \underbrace{x^2 + y^2}_{PO^2} + \underbrace{(x-1)^2 + y^2}_{PA^2} + \underbrace{(x-1)^2 + (y-1)^2}_{PB^2}$
$+ \underbrace{x^2 + (y-1)^2}_{PC^2}$

$= 4x^2 - 4x + 4y^2 - 4y + 4$
$= 4[x^2 - x + y^2 - y + 1]$
$= 4\left[\left(x - \dfrac{1}{2}\right)^2 - \dfrac{1}{4} + \left(y - \dfrac{1}{2}\right)^2 - \dfrac{1}{4} + 1\right]$
$= 4\left[\left(x - \dfrac{1}{2}\right)^2 + \left(y - \dfrac{1}{2}\right)^2 + \dfrac{1}{2}\right]$ as required

(a) since $(x-\frac{1}{2})^2 \geq 0$, $(y-\frac{1}{2})^2 \geq 0$

$d^2 \geq 4[0+0+\frac{1}{2}] = 2$, i.e. d^2 can't be less than 2

(b) $d^2 = 2 \Rightarrow 2 = 4[(x-\frac{1}{2})^2 + (y-\frac{1}{2})^2 + \frac{1}{2}]$

$\Rightarrow 2 = 4[(x-\frac{1}{2})^2 + (y-\frac{1}{2})^2] + 2$

$\Rightarrow (x-\frac{1}{2})^2 + (y-\frac{1}{2})^2 = 0$

i.e. $x = \frac{1}{2}$, $y = \frac{1}{2}$

(c) $4 = 4[(x-\frac{1}{2})^2 + (y-\frac{1}{2})^2]$

$(x-\frac{1}{2})^2 + (y-\frac{1}{2})^2 = 1$

i.e. circle, centre $(\frac{1}{2}, \frac{1}{2})$, radius 1

9 Mid-point of AB is $\left(\frac{-4+12}{2}, \frac{0+12}{2}\right)$ i.e. (4, 6)

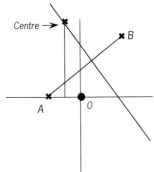

Gradient AB is $\frac{12}{16} = \frac{3}{4} \Rightarrow$ perpendicular gradient $\frac{-4}{3}$

$\Rightarrow$ equation is $y - 6 = -\frac{4}{3}(x-4)$

$3y - 18 = -4x + 16$

$3y + 4x = 34$

This intersects perpendicular bisector of AO, which is $x = -2$,

when $3y + 4(-2) = 34 \Rightarrow 3y = 42$, $y = 14$

i.e. centre has coordinates, (–2, 14)

$\Rightarrow (x+2)^2 + (y-14)^2 = 2^2 + 14^2 = 200$

10 (a) $q = -2$

(b) Substituting $y = 2$ when $x = 3$

$2 = \frac{p-6}{3-2} \Rightarrow p = 8$

(c) As $x \to \infty$, $y \to -2 \Rightarrow y = -2$ is asymptote

(d) $y = \frac{8-2x}{x-2}$

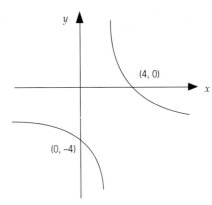

11 $y = \frac{2(x+1)+1}{x+1} = 2 + \frac{1}{x+1}$

i.e. translation, –1 x-direction
translation, 2 y-direction

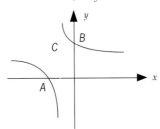

when $y = 0$, $2x + 3 = 0 \Rightarrow x = -\frac{3}{2}$, i.e. $A\left(\frac{-3}{2}, 0\right)$

$x = 0$, $y = 3$ i.e. $B(0, 3)$

the asymptotes are $x - 1$, $y = 2$, i.e. $C(-1, 2)$

12 $y = 1 - \frac{4}{x+2}$

i.e. translation –2 x-direction
stretch 4 y-direction
reflection x-axis
translation 1 y-direction

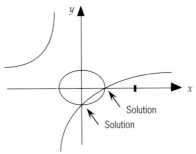

$y^2 = \frac{(x-2)^2}{(x+2)^2}$. If $\frac{x^2}{4} + y^2 = 1 \Rightarrow \frac{x^2}{4} + \frac{(x-2)^2}{(x+2)^2} = 1$

$\Rightarrow 4(x-2)^2 = (4-x^2)(x+2)^2$. Solutions (from sketch) $x = 2$, $x = 0$

Section 3

1 $(1+2x)^{\frac{1}{2}} = 1 + \frac{1}{2}(2x) + \frac{\frac{1}{2}(-\frac{1}{2})(2x)^2}{2!} +$

$\frac{\frac{1}{2}(-\frac{1}{2})(-\frac{3}{2})(2x)^3}{3!}$

$= 1 + x - \frac{1}{2}x^2 + \frac{1}{2}x^3$

2 $(4+x)^{\frac{-1}{2}} = [4(1+\frac{x}{4})]^{\frac{-1}{2}}$

$= 4^{\frac{-1}{2}}(1+\frac{x}{4})^{\frac{-1}{2}}$

$= \frac{1}{2}\left[1 + (-\frac{1}{2})(\frac{x}{4}) + \frac{(\frac{-1}{2})(\frac{-3}{2})(\frac{x}{4})^2}{2!}\right.$

$\left. + \frac{(\frac{-1}{2})(\frac{-3}{2})(\frac{-5}{2})(\frac{x}{4})^3}{3!}\right]$

$= \frac{1}{2} - \frac{1}{16}x + \frac{3x^2}{128} - \frac{5x^3}{1024}$

3 $(1+x)^{\frac{1}{3}} = 1 + \frac{1}{3}x + \frac{(\frac{1}{3})(\frac{-2}{3})x^2}{2!}$

$= 1 + \frac{1}{3}x - \frac{1}{9}x^2$

Putting $x = 0.0006$ gives

$(1.0006)^{\frac{1}{3}} = 1 + 0.0002 - 0.00000004$

$= 1.00019996$

4 $(1+x)^{-2} = 1 + (-2)(x) + \frac{(-2)(-3)(x)^2}{2!}$

$= 1 - 2x + 3x^2$

$\left(\frac{1-x}{1+x}\right)^2 = (1-x)^2(1+x)^{-2}$

$= (1-2x+x^2)(1-2x+3x^2)$

$= 1 - 2x - 2x + x^2 + 4x^2 + 3x^2$

$= 1 - 4x + 8x^2$ (ignoring x^3 and higher)

5 (a) $(1+12x)^{\frac{3}{4}} = 1 + (\frac{3}{4})(12x) + \frac{(\frac{3}{4})(-\frac{1}{4})(12x)^2}{2!}$

$+ \frac{(\frac{3}{4})(-\frac{1}{4})(-\frac{5}{4})(12x)^3}{3!}$

$= 1 + 9x - \frac{27}{2}x^2 + \frac{135}{2}x^3$

$\Rightarrow p = -\frac{27}{2}, q = \frac{135}{2}$

(b) Putting $x = 0.01$

$(1.12)^{\frac{3}{4}} \approx 1 + 0.09 - 0.0135 + 0.0000675$

$= 1.0887175$

$= 1.0887$ (4 d.p.)

6 (a) $(1+px)^{\frac{1}{3}} = 1 + \frac{1}{3}px + \frac{\frac{1}{3}(\frac{-2}{3})}{2}p^2x^2$

$= 1 + \frac{1}{3}px - \frac{1}{9}p^2x^2$

(b) $(1+qx)^{-1} = 1 - qx + \frac{(-1)(-2)}{2!}q^2x^2$

$= 1 - qx + e^2x^2$

$\Rightarrow \frac{1+2qx}{1+qx} = (1+2qx)(1-qx+q^2x^2)$

$= 1 + qx + (q^2 - 2q^2)x^2$

$= 1 + qx - q^2x^2$

If these are the same, $\frac{1}{3}p = q$ (coeffs of x)

$\Rightarrow p = 3q$ as required

7 (a) Repeated, so $\frac{A}{1+2x} + \frac{B}{1-x} + \frac{C}{(1+x)^2} \Rightarrow$

$5 + x = A(1-x)^2 + B(1+2x)(1-x) + C(1+2x)$

$x = 1$ $6 = 3C$ $\Rightarrow C = 2$

$x = \frac{-1}{2}$ $\frac{9}{2} = \frac{9}{4}A$ $\Rightarrow A = 2$

x^2 coeff $0 = A - 2B$ $\Rightarrow B = 1$

i.e. $\frac{2}{1+2x} + \frac{1}{1-x} + \frac{2}{(1-x)^2}$

(b) Rewritten, $2(1+2x)^{-1} + (1-x)^{-1} + 2(1-x)^{-2}$

$= 2[1 - 2x + 4x^2 - 8x^3] + [1 + x + x^2 + x^3] +$

$2[1 + 2x + 3x^2 + 4x^3]$

$= 5 + x + 15x^2 - 7x^3$

8 $(1+2x)^{\frac{-1}{2}} = 1 + (\frac{-1}{2})(2x) + \frac{(\frac{-1}{2})(\frac{-3}{2})}{2!}(2x)^2$

$+ \frac{(\frac{-1}{2})(\frac{-3}{2})(\frac{-5}{2})}{3!}(2x)^3$

$= 1 - x + \frac{3x^2}{2} - \frac{5}{2}x^3$

$\Rightarrow (1+5x)(1+2x)^{\frac{-1}{2}}$

$= 1 - x + \frac{3x^2}{2} - \frac{5}{2}x^3 + 5x - 5x^2 + \frac{15x^3}{2}$

$= 1 + 4x - \frac{7}{2}x^2 + 5x^3$

Putting $\lambda = 0.04$, $\frac{1+5x}{(1+2x)^{\frac{1}{2}}} = \frac{1.2}{(1.08)^{\frac{1}{2}}} = \frac{1.2}{(3 \times 0.36)^{\frac{1}{2}}}$

$= \frac{2}{\sqrt{3}}$

Into series, $1 + 0.16 - 0.0056 + 0.00032$

$= 1.15472$

$\Rightarrow \frac{1}{\sqrt{3}} = \frac{1.15472}{2} = 0.57736 = 0.577$ (3 d.p.)

Solutions P3

9 (a) $(8+x)^{\frac{1}{3}} = 8^{\frac{1}{3}}(1+\frac{x}{8})^{\frac{1}{3}}$

$$= 2\left[1 + \frac{x}{24} + \frac{(\frac{1}{3})(\frac{-2}{3})}{2!}(\frac{x}{8})^2\right]$$

$$= 2 + \frac{1}{12}x - \frac{1}{288}x^2$$

$$\Rightarrow p = \frac{1}{12}, \quad q = -\frac{1}{288}$$

(b) Putting $x = 7$, series gives 2.41 (to 3 s.f.)

(c) Percentage error is $\frac{\text{error}}{\text{exact}} \times 100 = 2.3\%$

10 $(1+2x)^{\frac{1}{2}} = 1 + x + \frac{(\frac{1}{2})(\frac{-1}{2})}{2}(2x)^2 = 1 + x - \frac{1}{2}x^2$

$(1+bx)^{-1} = 1 - bx + b^2x^2$

$\Rightarrow (1+ax)(1+bx)^{-1} = 1 - bx + b^2x^2 + ax - abx^2$

$= 1 + (a-b)x + (b^2 - ab)x^2$

If these are the same, $1 = a - b$

and $\frac{-1}{2} = b^2 - ab = b(b-a) = b(-1)$

$\Rightarrow b = \frac{1}{2}, \quad a = \frac{3}{2}$

Putting $x = -\frac{1}{100}$, $\sqrt{\frac{98}{100}} \approx \frac{1 - \frac{3}{200}}{1 - \frac{1}{200}} = \frac{197}{199}$

$\frac{7}{10}\sqrt{2} = \frac{197}{199} \Rightarrow \sqrt{2} \approx \frac{1970}{1393}$ [correct to 6 d.p.]

11 (a) Quadratic, so $\frac{A}{1-2x} + \frac{Bx+C}{1+x^2}$

$1 - 2x + 5x^2 \equiv A(1+x^2) + (Bx+C)(1-2x)$

$x = \frac{1}{2} \qquad \frac{5}{4} = \frac{5}{4}A \quad \Rightarrow A = 1$

$x = 0 \qquad 1 = A + C \quad \Rightarrow C = 0$

x^2 coeff $\quad 5 = A - 2B \quad \Rightarrow B = -2$

i.e. $\frac{1}{1-2x} - \frac{2x}{1+x^2}$

(b) $(1-2x)^{-1} = 1 + 2x + 4x^2 + 8x^3$

$(1+x^2)^{-1} = 1 - x^2$

$\Rightarrow \frac{1}{1-2x} - \frac{2x}{1+x^2} \approx 1 + 2x + 4x^2 + 8x^3 - 2x + 2x^3$

$= 1 + 4x^2 + 10x^3$

$a = 4, \quad b = 10$

12 (a) $f(x) \equiv \frac{9 - 3x - 12x^2}{(x-1)(2x+1)}$

$\Rightarrow 9 - 3x - 12x^2 \equiv A(1-x)(1+2x) + B(1+2x) + C(1-x)$

$x = 1 \qquad -6 = 3B \quad \Rightarrow B = -2$

$x = \frac{-1}{2} \qquad \frac{15}{2} = \frac{3C}{2} \quad \Rightarrow C = 5$

x^2 coeff $\quad -12 = -2A \quad \Rightarrow A = 6$

(b) $6 + \frac{5}{1+2x} - \frac{2}{1-x}$

$= 6 + 5(1 - 2x + 4x^2 - 8x^3) - 2(1 + x + x^2 + x^3)$

$= 9 - 12x + 18x^2 - 42x^3$

Section 4

1 $y = \dfrac{1-x}{1+x} \Rightarrow \dfrac{dy}{dx} = \dfrac{(1+x)(-1)-(1-x)(1)}{(1+x)^2}$

$= \dfrac{-1-x-1+x}{(1+x)^2} = \dfrac{-2}{(1+x)^2}$

$= -2(1+x)^{-2}$

$\dfrac{d^2y}{dx^2} = (-2)\left[(1-2)(1+x)^{-3}(1)\right]$

$= 4(1+x)^{-3} = \dfrac{4}{(1+x)^3}$

2 $y = \sin(x^3)$ $\dfrac{dy}{dx} = 3x^2 \cos(x^3)$ [Product]

$\dfrac{d^2y}{dx^2} = 3x^2\left[-3x^2 \sin(x^3)\right] + 6x \cos(x^3)$

$= 6x \cos(x^3) - 9x^4 \sin(x^3)$

3 Differentiating with respect to x,

$2e^{2x} + 2e^{2y}\dfrac{dy}{dx} = x\dfrac{dy}{dx} + y$

$\Rightarrow 2e^{2x} - y = x\dfrac{dy}{dx} - 2e^{2y}\dfrac{dy}{dx}$

$= (x - 2e^{2y})\dfrac{dy}{dx}$

$\Rightarrow \dfrac{dy}{dx} = \dfrac{2e^{2x} - y}{x - 2e^{2y}}$

4 $\dfrac{dx}{dt} = -2\sin t$: $\dfrac{dy}{dt} = -6\sin 2t$

$\dfrac{dy}{dx} = \dfrac{dy}{dt}\bigg/\dfrac{dx}{dt} = \dfrac{-6\sin 2t}{-2\sin t} = \dfrac{3\sin 2t}{\sin t}$

$= \dfrac{6\sin t \cos t}{\sin t} = 6\cos t$

Since $0 < t < \pi$, $\cos t \Rightarrow \dfrac{dy}{dx} < 6$

5 (a) If $x = 3$, $y^3 + y^2 + y = 3$

$y^3 + y^2 + y - 3 = (y-1)(y^2 + ay + 3) = 0$

y^2 coeff: $a - 1 = 1 \Rightarrow a = 2$

$\Rightarrow y^2 + 2y + 3 = 0$. Discriminant $\Delta = 2^2 - 12 < 0$

$\Rightarrow$ no further roots apart from 1, i.e.

the point (3, 1) is only intersection.

(b) Differentiating the original equation with respect to x,

$3y^2 \dfrac{dy}{dx} + 2y\dfrac{dy}{dx} + \dfrac{dy}{dx} = 2x - 2$

When $x = -1$, $y = 1$, $\dfrac{3dy}{dx} + \dfrac{2dy}{dx} + \dfrac{dy}{dx} = -4$

$\Rightarrow \dfrac{6dy}{dx} = -4$, $\dfrac{dy}{dx} = \dfrac{-2}{3}$

Equation tangent is $y - 1 = \dfrac{-2}{3}(x+1)$

$\Rightarrow 3y - 3 = -2(x+1) = -2x - 2$

$\Rightarrow 2x + 3y - 1 = 0$ as required

6 $f'(x) = \dfrac{(x^2+2)1 - x(2x)}{(x^2+2)^2} = \dfrac{2-x^2}{(x^2+2)^2}$

Since denominator $(x^2+2)^2 > 0$, we need $2 - x^2 < 0$

$x^2 > 2 \Rightarrow x < -\sqrt{2}$ or $x > \sqrt{2}$

7 (a) $y = \dfrac{\sin x}{x} \Rightarrow \dfrac{dy}{dx} = \dfrac{x\cos x - \sin x(1)}{x^2}$

$= \dfrac{x\cos x - \sin x}{x^2}$

(b) $y = \ln\left(\dfrac{1}{x^2+9}\right) = -\ln(x^2+9) \Rightarrow \dfrac{dy}{dx} = \dfrac{-2x}{x^2+9}$

(c) $\ln y = \ln x^x = x \ln x$

$\dfrac{1}{y}\dfrac{dy}{dx} = x\left(\dfrac{1}{x}\right) + \ln x = 1 + \ln x$

$\Rightarrow \dfrac{dy}{dx} = y(1 + \ln x) = x^x(1 + \ln x)$

8 (a) $\dfrac{dy}{dx} = \dfrac{2t}{3t^2} = \dfrac{2}{3t}$

When $x = 1$ and $y = 1$, $t = 1 \Rightarrow$ grad tangent is $\dfrac{2}{3}$

Equation is $y - 1 = \dfrac{2}{3}(x - 1) \Rightarrow 3y - 3 = 2x - 2$

i.e. $3y = 2x + 1$

(b) Substituting $x = t^3$, $y = t^2$ into l gives

$3t^2 - 2t^3 + 4 = 0$

$\Rightarrow 2t^3 - 3t^2 - 4 = 0$

Substituting $t = 2$ satisfies equation, then $t - 2$ is factor

$(t-2)(2t^2 + at + 2) = 2t^3 - 3t^2 - 4$

t^2 coeff $a - 4 = -3 \Rightarrow a = 1$

$\Rightarrow (t-2)(2t^2 + t + 2) = 0$

$2t^2 + t + 2 \neq 0$ since discriminant

$\Delta = 1 - 16 < 0$

$\Rightarrow t = 2 \Rightarrow$ coordinates of B are (8, 4) and this is the only intersection

9 (a) $f(x) = e^{2x}\sin 2x = 0 \Rightarrow \sin 2x = 0 \Rightarrow x = 0, \dfrac{\pi}{2}$ or π

(b) $f'(x) = e^{2x}(2\cos 2x) + 2e^{2x}\sin 2x$

$= 2e^{2x}(\cos 2x + \sin 2x)$

$2e^{2x} \neq 0 \Rightarrow \cos 2x + \sin 2x = 0 \Rightarrow \sin 2x = -\cos 2x$

$\tan 2x = -1$

$\Rightarrow 2x = \dfrac{3\pi}{4}, \dfrac{7\pi}{4}, \Rightarrow x = \dfrac{3\pi}{8}$ or $\dfrac{7\pi}{8}$

$x = \dfrac{3\pi}{8}, y = e^{\frac{3\pi}{4}}\sin^{\frac{3\pi}{4}} = \dfrac{1}{\sqrt{2}}e^{\frac{3\pi}{4}}$

$x = \dfrac{7\pi}{8}, y = e^{\frac{7\pi}{4}}\sin^{\frac{7\pi}{4}} = \dfrac{-1}{\sqrt{2}}e^{\frac{7\pi}{4}}$

(c) $f''(x) = 4e^{2x}(\cos 2x + \sin 2x)$
$\qquad\qquad + 2e^{2x}(-2\sin 2x + 2\cos 2x)$
$\qquad = 8e^{2x}\cos 2x$

(d) $x = \dfrac{3\pi}{8}, \quad f''(x) < 0$
$\qquad\qquad$ MAX

$\quad x = \dfrac{7\pi}{8}, \quad f''(x) > 0$
$\qquad\qquad$ MIN

10 (a) $\dfrac{dy}{dx} = \dfrac{1 + 2e^{2t}}{1 - 2e^{2t}}$

(b) If $\dfrac{dy}{dx} = 3 \Rightarrow \dfrac{1 + 2e^{2t}}{1 - 2e^{2t}} = 3 \Rightarrow 1 + 2e^{2t} = 3 - 6e^{2t}$

$\qquad\qquad 8e^{2t} = 2$

$\qquad\qquad e^{2t} = \dfrac{1}{4}$

Taking lns, $\quad 2t = \ln\dfrac{1}{4}$

$\qquad\qquad t = \dfrac{1}{2}\ln\dfrac{1}{4}$

$\qquad\qquad\quad = \dfrac{1}{2}[-2\ln 2] = -\ln 2$

11 $\dfrac{dx}{dt} = -3\cos^2 t\sin t \quad \dfrac{dy}{dt} = 3\sin^2 t\cos t$

$\dfrac{dy}{dx} = \dfrac{3\sin^2 t\cos t}{-3\cos^2 t\sin t} = -\dfrac{\sin t}{\cos t} = -\tan t$

Looking at y-coordinate,

$\sin^3 t = \dfrac{1}{8} \Rightarrow \sin t = \dfrac{1}{2} \Rightarrow t = \dfrac{\pi}{6}$

$\Rightarrow$ gradient of tangent is $-\tan\dfrac{\pi}{6} = -\dfrac{1}{\sqrt{3}}$

Equation of tangent is $y - \dfrac{1}{8} = \dfrac{-1}{\sqrt{3}}\left(x - \dfrac{3}{8}\sqrt{3}\right)$

$\Rightarrow y - \dfrac{1}{8} = \dfrac{-1}{\sqrt{3}}x + \dfrac{3}{8}$

$\Rightarrow y = \dfrac{-1}{\sqrt{3}}x + \dfrac{1}{2}$

$\qquad m = \dfrac{-1}{\sqrt{3}}$ and $c = \dfrac{1}{2}$.

12 $\dfrac{dy}{dx} = \dfrac{(2x-1)2x - x^2(2)}{(2x-1)^2} = \dfrac{4x^2 - 2x - 2x^2}{(2x-1)^2}$

$\qquad\quad = \dfrac{2x^2 - 2x}{(2x-1)^2} = \dfrac{2x(x-1)}{(2x-1)^2}$

If $\dfrac{dy}{dx} = 0 \Rightarrow 2x(x-1) = 0 \Rightarrow x = 0$ or $x = 1$

Section 5

1 (a) $\displaystyle\int_0^{\pi/4} \sec^2 x\,dx = \Big[\tan x\Big]_0^{\pi/4} = 1$

(b) Volume is $\pi\int y^2 dx = \pi\displaystyle\int_0^{\pi/4}\tan^2 x\,dx$

$\qquad\qquad = \pi\displaystyle\int_0^{\pi/4}(\sec^2 x - 1)\,dx$

$\qquad\qquad = \pi\Big[\tan x\Big]_0^{\pi/4}$

$\qquad\qquad = \pi\left(1 - \dfrac{\pi}{4}\right) = \pi - \dfrac{\pi^2}{4}$

2 Volume is $\pi\int y^2 dx = \pi\displaystyle\int_0^{\pi/4}\sin^2 x\,dx$

$\qquad\qquad = \dfrac{\pi}{2}\displaystyle\int_0^{\pi/4}(1 - \cos 2x)\,dx$

$\qquad\qquad = \dfrac{\pi}{2}\left[x - \dfrac{1}{2}\sin 2x\right]_0^{\pi/4}$

$\qquad\qquad = \dfrac{\pi}{2}\left[\dfrac{\pi}{4} - \dfrac{1}{2}\right] = \dfrac{\pi}{8}(\pi - 2)$

3 $\displaystyle\int_0^{\pi/4} x\sec^2 x\,dx \qquad u = x \quad \dfrac{dv}{dx} = \sec^2 x$

$\qquad\qquad\qquad\qquad\qquad \dfrac{du}{dx} = 1 \quad v = \tan x$

$= \Big[x\tan x - \textstyle\int\tan x\,dx\Big]_0^{\pi/4}$

$= \Big[x\tan x + \ln\cos x\Big]_0^{\pi/4}$

$= \left(\dfrac{\pi}{4}\times 1 + \ln\cos\dfrac{\pi}{4}\right) - (0 + \ln 1)$

$= \dfrac{\pi}{4} + \ln\dfrac{1}{\sqrt{2}} = \dfrac{\pi}{4} - \dfrac{1}{2}\ln 2$

$= \dfrac{1}{4}(\pi - \ln 4)$

4 $x = e^2, u = \ln e^2 = 2 \qquad \dfrac{du}{dx} = \dfrac{1}{x} \Rightarrow \dfrac{dx}{du} = x$
$\quad x = e, u = \ln e = 1$

$I = \displaystyle\int_1^2 \dfrac{1}{x\times u^{\frac{1}{2}}}\times x\,du = \displaystyle\int_0^2 u^{-\frac{1}{2}}\,du$

$\qquad\qquad = \left[2u^{\frac{1}{2}}\right]_1^2 = 2\sqrt{2} - 2$

5 Using $\sin 2x = 2\sin x\cos x$, $\dfrac{du}{dx} = \cos x \Rightarrow \dfrac{dx}{du} = \dfrac{1}{\cos x}$

$$I = \int \sin^3 x \times 2\sin x\cos x \times \dfrac{1}{\cos x}\, du$$
$$= \int 2\sin^4 x\, du = \int 2u^4\, du$$
$$= \dfrac{2}{5}u^5 + C$$
$$= \dfrac{2}{5}\sin^5 x + C$$

6 $u = 3 + e^{-x}$, $\dfrac{du}{dx} = -e^{-x}$, $x = 0, u = 4$
$x = \ln 0.5,\ u = 3 + e^{-\ln\frac{1}{2}} = 5$

$$I = \int_5^4 \dfrac{e^{-x}}{2u^{\frac{1}{2}}}\dfrac{dx}{du} \times du$$
$$= \int_5^4 \dfrac{e^{-x}}{2u^{\frac{1}{2}}} \times \dfrac{1}{-e^{-x}}\, du$$
$$= \int_5^4 -\dfrac{1}{2u^{\frac{1}{2}}}\, du = \left[u^{\frac{1}{2}}\right]_4^5 \quad \left(\begin{array}{c}\text{swapping limits}\\\text{and changing sign}\end{array}\right)$$
$$= (\sqrt{5} - 2)$$

7 $u = 4 + x^2$, $\dfrac{du}{dx} = 2x$, $x = 1, u = 5$; $x = 0, u = 4$

$$I = \int_4^5 \dfrac{x^3}{u^{\frac{1}{2}}}\dfrac{dx}{du} \times du = \int_4^5 \dfrac{x^3}{u^{\frac{1}{2}}} \times \dfrac{1}{2x}\, du$$
$$= \dfrac{1}{2}\int_4^5 \dfrac{x^2}{u^{\frac{1}{2}}}\, du$$

Put x^2 in terms of u
$$= \dfrac{1}{2}\int_4^5 \dfrac{u-4}{u^{\frac{1}{2}}}\, du$$
$$= \dfrac{1}{2}\int_4^5 u^{\frac{1}{2}} - 4u^{-\frac{1}{2}}\, du$$
$$= \dfrac{1}{2}\left[\dfrac{2}{3}u^{\frac{3}{2}} - 8u^{\frac{1}{2}}\right]_4^5$$
$$= \dfrac{1}{2}\left[(\dfrac{2}{3}\times 5\sqrt{5} - 8\sqrt{5}) - (\dfrac{2}{3}\times 8 - 8\times 3)\right]$$
$$= \dfrac{-7\sqrt{5}}{3} + \dfrac{16}{3} = \dfrac{1}{3}(16 - 7\sqrt{5})$$

8 From the factor formulae in P2,
$$2\cos A\cos B = \cos(A+B) + \cos(A-B)$$
Here $A + B = 6x$ and $A - B = 4x \Rightarrow A = 5x, B = x$
and $\cos 6x + \cos 4x = 2\cos 5x\cos x$, as required

$$\int_0^{\pi/12} \cos 5x\cos x\, dx = \dfrac{1}{2}\int_0^{\pi/12} (\cos 6x + \cos 4x)\, dx$$

$$= \dfrac{1}{2}\left[\dfrac{1}{6}\sin 6x + \dfrac{1}{4}\sin 4x\right]_0^{\pi/12}$$
$$= \dfrac{1}{2}\left[(\dfrac{1}{6} + \dfrac{\sqrt{3}}{8})\right]$$
$$= \dfrac{1}{48}(4 + 3\sqrt{3})$$

9 (a) $\dfrac{d(\ln(x^3 + 6x))}{dx} = \dfrac{3x^2 + 6}{x^3 + 6x} = \dfrac{3(x^2 + 2)}{x^3 + 6x}$

(b) $\displaystyle\int_2^3 \dfrac{x^2 + 2}{x^3 + 6x}\, dx = \left[\dfrac{1}{3}\ln(x^3 + 6x)\right]_2^3$
$$= \dfrac{1}{3}(\ln 45 - \ln 20)$$
$$= \dfrac{1}{3}\ln\left(\dfrac{45}{20}\right) = \dfrac{1}{3}\ln\dfrac{9}{4} = \dfrac{2}{3}\ln\dfrac{3}{2}$$

10 We need to integrate by parts twice
$\int x^2\sin 2x\, dx$ $u = x^2$ $V' = \sin 2x$
$u' = 2x$ $V = -\dfrac{1}{2}\cos 2x$

$$= -\dfrac{1}{2}x^2\cos 2x - \int 2x\left(-\dfrac{1}{2}\cos 2x\right) dx$$
$$= -\dfrac{1}{2}x^2\cos 2x + \int x\cos 2x\, dx \quad u = x\ \ V' = \cos 2x$$
$$u' = 1 \quad V = \dfrac{1}{2}\sin 2x$$
$$= -\dfrac{1}{2}x^2\cos 2x + \dfrac{x}{2}\sin 2x - \dfrac{1}{2}\int\sin 2x\, dx$$
$$= -\dfrac{1}{2}x^2\cos 2x + \dfrac{x}{2}\sin 2x + \dfrac{1}{4}\cos 2x + C$$

$$\int_0^{\pi/2} x^2\sin 2x$$
$$= \left[(-\dfrac{1}{2}(\dfrac{\pi}{2})^2\cos\pi + \dfrac{\pi}{4}\sin\pi + \dfrac{1}{4}\cos\pi) - (\dfrac{1}{4}\cos 0)\right]$$
$$= +\dfrac{\pi^2}{8} - \dfrac{1}{4} - \dfrac{1}{4} = \dfrac{1}{8}\pi^2 - \dfrac{1}{2}$$

$V = \pi\int y^2\, dx$
$$= \pi\int_0^{\pi/2} x^2(\cos x + \sin x)^2\, dx : (\cos x + \sin x)^2$$
$$= \cos^2 x + 2\sin x\cos x + \sin^2 x = 1 + 2\sin x\cos x$$
$$= 1 + \sin 2x$$
$$= \pi\int_0^{\pi/2} x^2(1 + \sin 2x)\, dx$$
$$= \pi\int_0^{\pi/2} x^2\, dx + \pi\int_0^{\pi/2} x^2\sin 2x\, dx$$
$$= \pi\left[\dfrac{x^3}{3}\right]_0^{\pi/2} + (\dfrac{1}{8}\pi^2 - \dfrac{1}{2})$$
$$= \dfrac{1}{24}\pi^4 + \dfrac{1}{8}\pi^2 - \dfrac{1}{2}$$

Solutions P3

11 (a) $\int \sin^2 x \, dx = \frac{1}{2}\int(1-\cos 2x)\, dx = \frac{1}{2}x - \frac{1}{4}\sin 2x + C$

(b) By parts, $\int_0^{\pi/2} x \sin^2 x \, dx$

$u = x \qquad V' = \sin^2 x$

$u' = 1 \qquad V = \frac{1}{2}x - \frac{1}{4}\sin 2x$ (from above)

$= \left[x\left(\frac{1}{2}x - \frac{1}{4}\sin 2x\right) - \int\left(\frac{1}{2}x - \frac{1}{4}\sin 2x\right) dx \right]_0^{\pi/2}$

$= \left[\frac{1}{2}x^2 - \frac{x}{4}\sin 2x - \left(\frac{1}{4}x^2 + \frac{1}{8}\cos 2x\right) \right]_0^{\pi/2}$

$= \left(\frac{1}{4}\left(\frac{\pi}{2}\right)^2 + \frac{1}{8}\right) - \left(-\frac{1}{8}\right)$

$= \frac{\pi^2}{16} + \frac{1}{4} = \frac{1}{16}(\pi^2 + 4)$

12 $\dfrac{6x+4}{(1-2x)(1+3x^2)} \equiv \dfrac{A}{1-2x} + \dfrac{Bx+C}{1+3x^2}$

$\Rightarrow 6x + 4 \equiv A(1+3x^2) + (1-2x)(Bx+C)$

$x = \frac{1}{2} \qquad 7 = A\left(\frac{7}{4}\right) \Rightarrow A = 4$

x^2-coeff $\quad 0 = 3A - 2B \Rightarrow B = 6$

constant $\quad 4 = A + C \Rightarrow C = 0$

i.e. $\dfrac{6x+4}{(1-2x)(1+3x^2)} = \dfrac{4}{1-2x} + \dfrac{6x}{1+3x^2}$

$\int_1^2 \dfrac{6x+4}{(1-2x)(1+3x^2)} dx$

$= \int_1^2 \left(\dfrac{4}{1-2x} + \dfrac{6x}{1+3x^2}\right) dx$

$= \left[-2\ln|1-2x| + \ln(1+3x^2)\right]_1^2$

$= (-2\ln 3 + \ln 13) - (-2\ln 1 + \ln 4)$

$= \ln 13 - \ln 4 - 2\ln 3$

$= \ln\left(\dfrac{13}{36}\right)$

13 (a) $\dfrac{5x^2 - 8x + 1}{2x(x-1)^2} \equiv \dfrac{A}{x} + \dfrac{B}{x-1} + \dfrac{C}{(x-1)^2}$

$\Rightarrow \dfrac{5x^2 - 8x + 1}{2} \equiv A(x-1)^2 + Bx(x-1) + Cx$

$x = 1 \qquad \dfrac{-2}{2} = C \Rightarrow C = \dfrac{-2}{2} = -1$

$x = 0 \qquad \dfrac{1}{2} = A$

x^2-coeff $\quad \dfrac{5}{2} = A + B \Rightarrow B = \dfrac{4}{2} = 2$

i.e. $A = \dfrac{1}{2}, B = 2, C = -1$

(b) $\int f(x)\, dx = \int \dfrac{1}{2x} + \dfrac{2}{x-1} - \dfrac{1}{(x-1)^2} \cdot dx$

$= \dfrac{1}{2}\ln(x) + 2\ln(x-1) + \dfrac{1}{x-1} + C$

(c) $\int_1^2 f(x)\, dx$

$= \left[\left(\dfrac{1}{2}\ln 9 + 2\ln 8 + \dfrac{1}{8}\right) - \left(\dfrac{1}{2}\ln 4 + 2\ln 3 + \dfrac{1}{3}\right)\right]$

$= \dfrac{1}{2}\ln 9 - \dfrac{1}{2}\ln 4 + 2\ln 8 - 2\ln 3 - \dfrac{5}{24}$

$= \ln 3 - \ln 2 + 2\ln\dfrac{8}{3} - \dfrac{5}{24}$

$= \ln\dfrac{3}{2} + \ln\dfrac{64}{9} - \dfrac{5}{24}$

$= \ln\dfrac{32}{3} - \dfrac{5}{24} \qquad$ as required

14 (a) $\int x(x^2+3)^5 dx$

This comes from $\dfrac{d(x^2+3)^6}{dx} = 6(x^2+3) \times 2x$

$= 12x(x^2+3)^5$

$I = \dfrac{1}{12}(x^2+3)^6 + C$

(b) By parts $\int_1^e \dfrac{1}{x^2}\ln x \, dx \quad u = \ln x \quad V' = \dfrac{1}{x^2}$

$u' = \dfrac{1}{x} \qquad V = -\dfrac{1}{x}$

$= \left[-\dfrac{1}{x}\ln x - \int \dfrac{1}{x} \times -\dfrac{1}{x} dx\right]_1^e$

$= \left[-\dfrac{1}{x}\ln x + \int \dfrac{1}{x^2} dx\right]_1^e$

$= \left[-\dfrac{1}{x}\ln x - \dfrac{1}{x}\right]_1^e$

$= \left(-\dfrac{1}{e}\ln e - \dfrac{1}{e}\right) - \left(-\ln 1 - 1\right) = 1 - \dfrac{2}{e}$

as required

(c) By parts $\int_1^p \dfrac{1}{(x+1)(2x-1)} dx$

$= \int_1^p \left[-\dfrac{1}{3(x+1)} + \dfrac{2}{3(2x-1)}\right] dx$

$= \dfrac{1}{3}\left[-\ln(x+1) + \ln(2x-1)\right]_1^p$

$= \dfrac{1}{3}\left[-\ln(p+1) + \ln(2p-1) + \ln 2\right]$

$= \dfrac{1}{3}\ln\dfrac{4p-2}{p+1}$

145

15 (a) $\dfrac{dx}{dt} = 2(t+1)$, $\dfrac{dy}{dt} = \dfrac{3}{2}t^2$

$\dfrac{dy}{dx} = \dfrac{dy}{dt} \bigg/ \dfrac{dx}{dt} = \dfrac{3t^2}{4(t+1)}$

When $t = 2$, $\dfrac{dy}{dx} = \dfrac{12}{4(3)} = 1$; $x = 9$, $y = 7$

Grad. normal is $-1 \Rightarrow$ equation is $y - 7 = -(x - 9)$

$x + y = 16$

(b) Area can be split into two

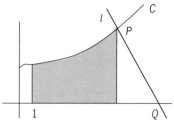

When $t = 2$, $x = 9$ and $y = 7$. Normal crosses the x-axis at the point Q where $y = 0 \Rightarrow x = 16$.

Area of Δ is $\dfrac{1}{2}$ base $\times$ height $= \dfrac{1}{2} \times (16 - 9) \times 7$

$= \dfrac{49}{2}$

The remaining area is from $t = 2$ (at P) to $t = 0$ (at $x = 1$)

$\dfrac{dx}{dt} = 2t + 2 \Rightarrow$ area is $\displaystyle\int_0^2 \left(\dfrac{1}{2}t^3 + 3\right)(2t + 2)\, dt$

$= \displaystyle\int_0^2 (t^4 + t^3 + 6t + 6)\, dt$

$= \left[\dfrac{t^5}{5} + \dfrac{t^4}{4} + 3t^2 + 6t\right]_0^2$

$= \dfrac{32}{5} + 4 + 12 + 12 = \dfrac{172}{5}$

$\Rightarrow$ total shaded area is $\dfrac{49}{2} + \dfrac{172}{5} = 58.9$

16 (a) $\int xe^{2x} dx$ by parts $u = x$ $V' = e^{2x}$
$\qquad\qquad\qquad\qquad\quad u' = 1$ $V = \dfrac{1}{2}e^{2x}$

$= \dfrac{x}{2}e^{2x} - \int \dfrac{1}{2} e^{2x} dx$

$= \dfrac{x}{2} e^{2x} - \dfrac{1}{4} e^{2x} + C$

$R_1: \left[\dfrac{x}{2} e^{2x} - \dfrac{1}{4} e^{2x}\right]_{-\frac{1}{2}}^{0}$

$= -\dfrac{1}{4} - \left[-\dfrac{1}{4}e^{-1} - \dfrac{1}{4}e^{-1}\right]$

$= \dfrac{1}{2e} - \dfrac{1}{4}$

This is negative, since under the axis, so $A_1 = \dfrac{1}{4} - \dfrac{1}{2e}$

$R_2: \left[\dfrac{x}{2}e^{2x} - \dfrac{1}{4}e^{2x}\right]_0^{\frac{1}{2}} = \left(\dfrac{1}{4}e - \dfrac{1}{4}e\right) - \left(-\dfrac{1}{4}\right) = \dfrac{1}{4}$

So $A_1 = \dfrac{1}{4} - \dfrac{1}{2e}$, $A_2 = \dfrac{1}{4}$

(b) $\dfrac{A_1}{A_2} = \dfrac{\frac{1}{4} - \frac{1}{2e}}{\frac{1}{4}} = \dfrac{1 - \frac{2}{e}}{1} = \dfrac{e-2}{e}$ as required

17 (a) $\dfrac{dy}{d\theta} = 4\cos\theta$ $\dfrac{dx}{d\theta} = -5\sin\theta$

$\dfrac{dy}{dx} = -\dfrac{4\cos\theta}{5\sin\theta}$, when $\theta = \dfrac{\pi}{4}$, grad is $\dfrac{-4}{5}$

(b) When $\theta = \dfrac{\pi}{4}$, $x = \dfrac{5}{\sqrt{2}}$ and $y = \dfrac{4}{\sqrt{2}}$

Equation tangent is $y - \dfrac{4}{\sqrt{2}} = \dfrac{-4}{5}\left(x - \dfrac{5}{\sqrt{2}}\right)$

$\times$ by 5. $5y - \dfrac{20}{\sqrt{2}} = -4x + \dfrac{20}{\sqrt{2}}$

$\Rightarrow 4x + 5y = \dfrac{40}{\sqrt{2}} = 20\sqrt{2}$

(c) When $y = 0$, $x = 5\sqrt{2} \Rightarrow R(5\sqrt{2}, 0)$

(d) Drop a perpendicular from P onto x-axis

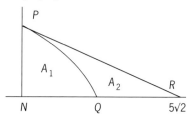

At Q, $y = 0 \Rightarrow 4\sin\theta = 0$ $\theta = 0 \Rightarrow x = 5$

At N, $\theta = \dfrac{\pi}{4}$, $x = \dfrac{5}{\sqrt{2}}$

For ΔPNR, which is $A_1 + A_2$, $\dfrac{1}{2} \times NR \times PN$

$= \dfrac{1}{2} \times \left(5\sqrt{2} - \dfrac{5}{\sqrt{2}}\right) \times \dfrac{4}{\sqrt{2}}$

$= 10 - 5 = 5$

For area A_1, from $\theta = 0$ at Q to $\theta = \dfrac{\pi}{4}$ at P,

$\dfrac{dx}{d\theta} = -5\sin\theta \Rightarrow$ area is $\displaystyle\int_{\pi/4}^{0} 4\sin\theta(-5\sin\theta)\, d\theta$

swapping limits, changing sign $= \displaystyle\int_0^{\pi/4} 20\sin^2\theta\, d\theta$

$= 10 \displaystyle\int_0^{\pi/4} (1 - \cos 2\theta)\, d\theta$

$$= \left[10\theta - 5\sin 2\theta\right]_0^{\pi/4}$$

$$= \frac{10\pi}{4} - 5 = \frac{5\pi}{2} - 5$$

$$\Rightarrow A_2 = \Delta - A_1 = 5 - \left(\frac{5\pi}{2} - 5\right) = 10 - \frac{5\pi}{2}$$

Section 6

1 $\frac{dy}{dx} = \frac{2y}{x} \Rightarrow \int \frac{1}{y}dy = \int \frac{2}{x}dx$

$\ln y = 2\ln x + C$
$= \ln x^2 + C = \ln Ax^2 \;(C = \ln A)$

Taking e's, $y = Ax^2$

If $x = 2, y = 2, y = \frac{1}{2}x^2$

If $x = -1, y = -1, y = -x^2$

with graphs

2 Rearranging, $\int \frac{2y}{1+y^2}dy = \int \frac{1+x^2}{x^2}dx$

$$= \int \left(\frac{1}{x^2} + 1\right)dx$$

$$\Rightarrow \ln(1+y^2) = \frac{-1}{x} + x + C$$

when $y = 1, x = 1,$ $\ln 2 = -1 + 1 + C \Rightarrow C = \ln 2$

$$\ln(1+y^2) - \ln 2 = \frac{-1}{x} + x$$

$$\ln\frac{1+y^2}{2} = \frac{x^2 - 1}{x}$$

Taking e's $\frac{1+y^2}{2} = e^{(x^2-1)/x}$

$1 + y^2 = 2e^{(x^2-1)/x}$

$y^2 = 2e^{(x^2-1)/x} - 1$

3 OP has gradient $\frac{y}{x}$

AP has gradient $\frac{y}{x-2}$

Gradient of tangent is $\frac{dy}{dx}$

$\Rightarrow$ gradient of normal is $\frac{-dx}{dy}$

$\Rightarrow \frac{-dx}{dy} = \left(\frac{y}{x}\right)\left(\frac{y}{x-2}\right)$

$\Rightarrow \frac{dy}{dx} = \frac{2x - x^2}{y^2}$

Rearranging $\int y^2 dy = \int (2x - x^2)dx$

$\frac{y^3}{3} = x^2 - \frac{1}{3}x^3 + C$

$x = 1, y = 1$ $\frac{1}{3} = 1 - \frac{1}{3} + C \Rightarrow C = -\frac{1}{3}$

$\Rightarrow y^3 = 3x^2 - x^3 - 1$

4 (a) $\int xe^{-x}dx$ $\quad u = x \quad V' = e^{-x}$
$\quad u' = 1 \quad V = -e^{-x}$

$= -xe^{-x} - \int -e^{-x}dx$

$= -xe^{-x} - e^{-x} + C$

(b) Rearranging, $\int \sin 2y \, dy = \int xe^{-x} dx$

$-\frac{1}{2}\cos 2y = -xe^{-x} - e^{-x} + C$

When $y = \frac{\pi}{4}$, $x = 0$ $\quad 0 = -1 + C \Rightarrow C = 1$

$\Rightarrow \cos 2y = 2e^{-x}(x+1) - 2$

5 $\frac{dr}{dt} = \frac{k}{r}$: when $r = 2$, $\frac{dr}{dt} = 0.1$

$\Rightarrow 0.1 = \frac{k}{2}$, $k = 0.2 \Rightarrow \frac{dr}{dt} = \frac{0.2}{r}$

i.e. $r\frac{dr}{dt} = 0.2$

Rearranging, $\int r \, dr = \int 0.2 \, dt$

$\frac{r^2}{2} = 0.2t + C$ ... ①

$t = 0 \Rightarrow A = 4\pi \Rightarrow \pi r^2 = 4\pi \Rightarrow r = 2$

Into ①, $2 = C \Rightarrow \frac{r^2}{2} = 0.2t + 2$ ... ②

$A = 16\pi \Rightarrow \pi r^2 = 16\pi$, $r = 4$

Into ②, $\frac{4^2}{2} = 0.2t + 2 \Rightarrow t = 30$ seconds.

6 (a) $u = 1 + 2x \Rightarrow \frac{du}{dx} = 2$: $2x = u - 1 \Rightarrow 4x = 2(u-1)$

$\int \frac{4x}{u^2} \times \frac{du}{2} = \int \frac{2(u-1)}{u^2} \times \frac{du}{2}$

$= \int \frac{u-1}{u^2} du$

$= \int \left(\frac{1}{u} - \frac{1}{u^2}\right) du$

$= \ln u + \frac{1}{u} + C$

$= \ln(1+2x) + \frac{1}{1+2x} + C$

(b) $\int \sin^2 y \, dy = \int \frac{x}{(1+2x)^2} dx$

$\int \sin^2 y \, dy = \frac{1}{2}\int (1 - \cos 2y) dy$

$= \frac{1}{2}\left(y - \frac{1}{2}\sin 2y\right)$

$= \frac{1}{4}\left(\ln(1+2x) + \frac{1}{1+2x}\right) + C$

$y = \frac{\pi}{4}$ when $x = 0$, $\frac{1}{2}\left(\frac{\pi}{4} - \frac{1}{2}\right) = \frac{1}{4}(\ln 1 + 1) + C$

$\Rightarrow C = \frac{\pi}{8} - \frac{1}{2}$

Multiply by 4, $2y - \sin 2y$

$= \ln(1+2x) + \frac{1}{1+2x} + \frac{\pi}{2} - 2$

7 (a) $\frac{dV}{dt} = -kV \Rightarrow \int \frac{1}{V} dv = \int -k \, dt$

$\ln V = -kt + C$

$V = e^{-kt+C} = Ae^{-kt}$ $(A = e^C)$

(b)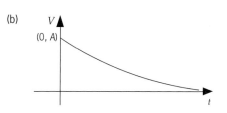

(c) Substituting, $\frac{1}{2}A = Ae^{-kT}$

$\frac{1}{2} = e^{-kT}$

$\ln \frac{1}{2} = -kT$

$\Rightarrow kT = -\ln\frac{1}{2} = \ln 2$

8 (a) Separating, $\int \frac{1}{P} dP = \int k \cos kt \, dt$

$\Rightarrow \ln P = \sin kt + C$ ... ①

$P = P_0$ when $t = 0$ $\ln P_0 = 0 + C \Rightarrow C = \ln P_0$

Into ① gives $\ln P = \sin kt + \ln P_0$

$\ln P - \ln P_0 = \sin kt$

$\ln \frac{P}{P_0} = \sin kt$

$\frac{P}{P_0} = e^{\sin kt}$

$P = P_0 e^{\sin kt}$

(b) Max of sine is 1 $\Rightarrow P_{Max} = P_0 e^1$

Min of sine is -1 $\Rightarrow P_{Min} = P_0 e^{-1}$

$\frac{P_{Max}}{P_{Min}} = \frac{P_0 e}{P_0 e^{-1}} = e^2 : 1$

9 (a) $\frac{dx}{dt} = -kx \Rightarrow \int \frac{1}{x} dx = \int -k \, dt$

$\ln x = -kt + C$: $x = A$ when $t = 0 \Rightarrow C = \ln A$

$\ln x = -kt + \ln A$

$\ln x - \ln A = -kt$

$\ln \frac{x}{A} = -kt \Rightarrow \frac{x}{A} = e^{-kt}$, $x = Ae^{-kt}$

(b) Substituting values,

$\frac{1}{3}A = Ae^{-10k}$

$\Rightarrow \frac{1}{3} = e^{-10k}$, $\ln\frac{1}{3} = -10k$ taking lns

$k = -\frac{1}{10}\ln\frac{1}{3}$

$\Rightarrow$ putting this and $x = \frac{1}{2}A$ into $\ln\frac{x}{A} = -kt$ gives

$\ln\frac{1}{2} = -\left(\frac{-1}{10}\ln\frac{1}{3}\right)t$

$\Rightarrow t = \frac{+10 \ln \frac{1}{2}}{\ln \frac{1}{3}} = 6.31$ (2 d.p.)

10 $\dfrac{dn}{dt} = kn \Rightarrow \int \dfrac{1}{n}\,dn = \int k\,dt$

$\ln n = kt + C$

$n = e^{kt+C} = Ae^{kt}$ where $A = e^C$

Take the *whole* RHS into the LHS

$\int \dfrac{1}{kn-p}\,du = \int dt$

$\dfrac{1}{k}\ln|kn-p| = t + C$ $t = 0, n = 500, k = 2, p = 100$

$\dfrac{1}{2}\ln|1000 - 100| = C = \dfrac{1}{2}\ln 900$

$\Rightarrow \dfrac{1}{2}\ln(2n - 100) - \dfrac{1}{2}\ln 900 = t$

$\ln\dfrac{2n-100}{900} = 2t \Rightarrow \dfrac{2n-100}{900} = e^{2t}$

$2n - 100 = 900e^{2t} \Rightarrow n = 50(1 + 9e^{2t})$

As it stands, the number will increase indefinitely (and very quickly) – not a realistic model beyond a certain point.

11 (a) Volume is base area × height $\Rightarrow V = 8x$

$\dfrac{dV}{dx} = 8$ ... ①

Two factors, increase in $\dfrac{dV}{dt}$ of 0.01

decrease in $\dfrac{dV}{dt}$ of $\dfrac{1}{4}x^2$

$\Rightarrow \dfrac{dV}{dt} = 0.01 - \dfrac{1}{4}x^2$

$\dfrac{dV}{dt} = \dfrac{dV}{dx} \times \dfrac{dx}{dt} = \dfrac{8\,dx}{dt} = 0.01 - \dfrac{1}{4}x^2$ from ①

i.e. $\dfrac{dx}{dt} = \dfrac{1}{800} - \dfrac{1}{32}x^2$, $\lambda = \dfrac{1}{800}$, $\mu = \dfrac{1}{32}$

(b) Rearranging, $\int \dfrac{800}{1-25x^2}\,dx = \int dt$

Partial fractions, $\dfrac{800}{1-25x^2} = \dfrac{400}{1-5x} + \dfrac{400}{1+5x}$

$\Rightarrow 80[\ln(1+5x) - \ln(1-5x)] = t + C$

$80\ln\left(\dfrac{1+5x}{1-5x}\right) = t + C$

$t = 0, x = 0 \Rightarrow C = 0$

$80\ln\left(\dfrac{1+5x}{1-5x}\right) = t$

when $x = 0.1$, $t = 80\ln\dfrac{1.5}{0.5} = 80\ln 3$ seconds.

12 (a) $\dfrac{1}{(3x-1)(x)} = \dfrac{3}{3x-1} - \dfrac{1}{x}$

(b) Rearranging,

$\int \dfrac{1}{3P^2 - P}\,dP = \int \dfrac{1}{2}\sin t\,dt$

$\int \left(\dfrac{3}{3P-1} - \dfrac{1}{P}\right)dP = \int \dfrac{1}{2}\sin t\,dt$

$\ln(3P-1) - \ln P = \dfrac{-1}{2}\cos t + C$

$P = \dfrac{1}{2}, t = 0$

$\ln\dfrac{1}{2} - \ln\dfrac{1}{2} = -\dfrac{1}{2} + C \Rightarrow C = \dfrac{1}{2}$

$\ln(3P-1) - \ln P = \dfrac{1}{2} - \dfrac{1}{2}\cos t$

i.e. $\ln\left(\dfrac{3P-1}{P}\right) = \dfrac{1}{2}(1 - \cos t)$ ... ①

as required

(c) When $P = 1$, $\ln 2 = \dfrac{1}{2}(1 - \cos t)$

$1 - \cos t = 2\ln 2$

$\cos t = 1 - 2\ln 2$

$\Rightarrow t = \text{inv}\cos(1 - 2\ln 2)$

$= 1.97$ rads (2 d.p.)

(d) Taking e's of both sides of ①,

$\dfrac{3P-1}{P} = e^{\frac{1}{2}(1-\cos t)}$

$\Rightarrow 3P - 1 = Pe^{\frac{1}{2}(1-\cos t)}$

$3P - Pe^{\frac{1}{2}(1-\cos t)} = 1$

$P\left(3 - e^{\frac{1}{2}(1-\cos t)}\right) = 1$

$\Rightarrow P = \dfrac{1}{3 - e^{\frac{1}{2}(1-\cos t)}}$

Minimum and maximum values of $\cos t$ are -1 and 1, giving corresponding values for P of

$\dfrac{1}{3-e^1} = \dfrac{1}{3-e}$ (Max) and $\dfrac{1}{3-e^0} = \dfrac{1}{2}$ (Min)

13 (a) Rearranging, $\int \dfrac{1}{P}\,dP = \int k\,dt \Rightarrow \ln P = kt + C$

$t = 0, P = P_0 \Rightarrow C = \ln P_0 \Rightarrow \ln P - \ln P_0 = kt$

$\ln\dfrac{P}{P_0} = kt$

$\dfrac{P}{P_0} = e^{kt}$ i.e. $P = P_0 e^{kt}$

(b) $P_0 = 1218 \Rightarrow P = 1218 e^{kt}$

$t = 1, P = 1397 \Rightarrow 1397 = 1218 e^k \Rightarrow e^k = \dfrac{1397}{1218}$

$t = 2$, $P = 1218 e^{2k} = 1218\left(\dfrac{1397}{1218}\right)^2 = 1602$

(c) Increases indefinitely

Rearranging, $\int \dfrac{1}{P}\,dP = \int ke^{-\lambda t}\,dt$

$\ln P = -\dfrac{1}{\lambda}ke^{-\lambda t} + C$ ... ①

$t = 0, P = P_0 \Rightarrow \ln P_0 = -\dfrac{1}{\lambda}k + C$

$\Rightarrow C = \ln P_0 + \dfrac{1}{\lambda}k$

Back into ①, $\ln P - \ln P_0 = \dfrac{1}{\lambda}k - \dfrac{1}{k}ke^{-\lambda t}$

$\ln \dfrac{P}{P_0} = \dfrac{1}{\lambda}(1 - e^{-\lambda t})$

P3 Solutions

As $t \to \infty$, $e^{-\lambda t} \to 0 \Rightarrow \ln \dfrac{P}{P_0} \to \dfrac{k}{\lambda}$

$\Rightarrow P \to P_0 e^{k/\lambda}$

14 (a) Falls 80° in 20 mins ⇒ to fall 120° in 30 mins, i.e. at time 6.30 pm.

(b) $\dfrac{d\theta}{dt} = -k(\theta - 20)$

$\int \dfrac{1}{\theta - 20} d\theta = \int -k\, dt \quad \ln|\theta - 20| = -kt + C$

$t = 0, \theta = 150 \Rightarrow C = \ln 130$

$kt = \ln 130 - \ln|\theta - 20|$

When $t = 20, \theta = 70 \Rightarrow 20k = \ln \dfrac{130}{50} = \ln \dfrac{13}{5}$

$k = \dfrac{1}{20} \ln \dfrac{13}{5}$

$\Rightarrow t = \dfrac{20 \ln 130}{\ln \dfrac{13}{5}} - \dfrac{20}{\ln \dfrac{13}{5}} \ln|\theta - 20|$

which is the required form, with $a = \dfrac{-20}{\ln \dfrac{13}{5}}$,

$b = \dfrac{20 \ln 130}{\ln \dfrac{13}{5}}$

when $\theta = 30°$, $t = \dfrac{20 \ln 130}{\ln \dfrac{13}{5}} - \dfrac{20}{\ln \dfrac{13}{5}} \ln 10$

$= \dfrac{20}{\ln \dfrac{13}{5}} [\ln 130 - \ln 10]$

$= \dfrac{20 \ln 13}{\ln \dfrac{13}{5}} = 54$ (nearest)

i.e. time will be 6.54 pm.

15 (a) If h is height, Volume $V = h \times 100^2 = h \times 10^4$ cm^3

$\dfrac{dV}{dt}$ has two components:

outflow (negative) of − 900

inflow (positive) of 400

$\Rightarrow \dfrac{dV}{dt} = 400 - 900 = -500$

Initially, $V = 81 \times 10^4 \Rightarrow$ when $h = 64, V = 64 \times 10^4$

$\Rightarrow$ Difference is 17×10^4 at rate of 500 cm^3 s^{-1}

$\Rightarrow$ time is 340 s

(b) (i) Loss $\dfrac{dV}{dt} = -k\sqrt{h}$ since $V = h \times 10^4$, $\dfrac{dV}{dh} = 10^4$

$\Rightarrow \dfrac{dh}{dt} = \dfrac{dh}{dV} \times \dfrac{dV}{dt} = \dfrac{1}{10^4} \times -k\sqrt{h}$

Since $\dfrac{dV}{dt} = -900$ when $h = 81$,

$-k\sqrt{81} = -900$

$\Rightarrow k = 100$

$\Rightarrow \dfrac{dh}{dt} = \dfrac{-\sqrt{h}}{100}$. This is loss.

Gain $\dfrac{dV}{dt} = 400 = 10^4 \dfrac{dh}{dt} \Rightarrow \dfrac{dh}{dt} = \dfrac{400}{10^4}$

This is gain.

Altogether, $\dfrac{dh}{dt} = \dfrac{400}{10^4} - \dfrac{\sqrt{h}}{100} = 0.04 - 0.01\sqrt{h}$

(ii) Rearranging $\int \dfrac{1}{0.04 - 0.01\sqrt{h}} dh = \int dt$

i.e. $\int \dfrac{100}{4 - \sqrt{h}} dh = t$

$x = \sqrt{h} - 4 \Rightarrow \dfrac{dx}{dh} = \dfrac{1}{2\sqrt{h}} \Rightarrow$

$t = \int \dfrac{100}{-x} \times 2\sqrt{h}\, dx = \int \dfrac{200(x+4)}{-x} dx$

$= -200 \int \left(1 + \dfrac{4}{x}\right) dx$

$\Rightarrow t = -200[x + 4\ln x] + C$

$t = 0, x = \sqrt{81} - 4 = 5 \Rightarrow C = 200(5 + 4\ln 5)$

$h = 64 \Rightarrow x = \sqrt{64} - 4 = 4$

$\Rightarrow t = 200(5 + 4\ln 5) - 200(4 + 4\ln 4)$

$= 200 + 800 \ln \dfrac{5}{4} = 379$ s (nearest)

16 (a) If $\dfrac{dT}{dx}$ has the same value, it is given as −6,

$\dfrac{dT}{dx} = -6 \Rightarrow T = -6x + C$

when $x = 30, T = 290$: $290 = -180 + C \Rightarrow C = 470$

i.e. $T = 470 - 6x$, when $x = 60, T = 110°$

i.e. temperature difference of 360°.

(b) $\dfrac{dT}{dx} = kx \Rightarrow T = \dfrac{kx^2}{2} + C$

Since $\dfrac{dT}{dx} = -6$ when $x = 30$,

$-6 = 30k \Rightarrow k = \dfrac{-1}{5} \Rightarrow T = \dfrac{-x^2}{10} + C$

when $x = 30, T = 290$

$290 = -\dfrac{900}{10} + C \Rightarrow C = 380$

$\Rightarrow T = 380 - \dfrac{x^2}{10}$

When $x = 0, T = 380$

$x = 60, T = 380 - \dfrac{3600}{10} = 20$

17 (a) $A = \pi r^2 \Rightarrow \dfrac{dA}{dr} = 2\pi r \Rightarrow \dfrac{dr}{dA} = 1 / \dfrac{dA}{dr} = \dfrac{1}{2\pi r}$

(b) $\dfrac{dr}{dt} = \dfrac{dr}{dA} \times \dfrac{dA}{dt} = \dfrac{1}{2\pi r} \times \dfrac{dA}{dt} = \dfrac{1}{2\pi r} \dfrac{2}{(t+1)^3}$

$\dfrac{dr}{dt} = \dfrac{1}{\pi r(t+1)^3}$

(c) Separating, $\int \pi r\, dr = \int \dfrac{1}{(t+1)^3} dt$

$\Rightarrow \dfrac{\pi r^2}{2} = -\dfrac{1}{2(t+1)^2} + C$

$A = 0 \Rightarrow r = 0$ when $t = 0 \Rightarrow$

$$0 = \frac{-1}{2} + C, \quad C = \frac{1}{2}$$

$$\Rightarrow \frac{\pi r^2}{2} = \frac{1}{2} - \frac{1}{2(t+1)^2}$$

Since $A = \pi r^2$, $\frac{A}{2} = \frac{1}{2} - \frac{1}{2(t+1)^2}$

$$\Rightarrow A = 1 - \frac{1}{(t+1)^2}$$

(d) $t = 1$, $A = \frac{3}{4} \Rightarrow \pi r^2 = \frac{3}{4}$, $r = \frac{\sqrt{3}}{2\sqrt{\pi}}$

$$\Rightarrow \frac{dr}{dt} = \frac{1}{\pi \times \frac{\sqrt{3}}{2\sqrt{\pi}}(2)^3} = \frac{1}{4\sqrt{3\pi}} = 0.081 \text{ (2 sf's)}$$

18 (a) $\int \frac{1}{V} dV = \int \frac{-1}{10} dt \Rightarrow \ln V = \frac{-t}{10} + C$, $t = 0$, $V = 5$

$$\Rightarrow C = \ln 5, \quad \ln V - \ln 5 = \frac{-t}{10}$$

$$\ln \frac{V}{5} = \frac{-t}{10} \quad \frac{V}{5} = e^{\frac{-t}{10}} \Rightarrow V = 5e^{\frac{-t}{10}}$$

(i) When $t = 1$, $V = 5e^{\frac{-1}{10}} = 4.52$ (3 sf's)

(ii) Infinite lifetime: $e^{\frac{-t}{10}} \neq 0$

(b) $\frac{dr}{dt} = -k \Rightarrow r = -kt + C$

Sphere $\Rightarrow V = \frac{4}{3}\pi r^3$

When $t = 0$, $V = 5 \Rightarrow \frac{4}{3}\pi r^3 = 5$

$$\Rightarrow r^3 = \frac{15}{4\pi}$$

$$\Rightarrow C = \sqrt[3]{\frac{15}{4\pi}} \text{ and } r = \sqrt[3]{\frac{15}{4\pi}} - kt \quad \ldots \text{①}$$

When $t = 1$, $V = \frac{4}{3}\pi r^3 = \frac{9}{2} \Rightarrow r^3 = \frac{27}{8\pi} \Rightarrow r = \sqrt[3]{\frac{27}{8\pi}}$

Into ①, $\sqrt[3]{\frac{27}{8\pi}} = \sqrt[3]{\frac{15}{4\pi}} - k \Rightarrow k = \sqrt[3]{\frac{15}{4\pi}} - \sqrt[3]{\frac{27}{8\pi}}$

From ①, when $r = 0$, $t = \frac{1}{k}\sqrt[3]{\frac{15}{4\pi}} = \frac{\sqrt[3]{\frac{15}{4\pi}}}{\sqrt[3]{\frac{15}{4\pi}} - \sqrt[3]{\frac{27}{8\pi}}}$

$= 29.0$ weeks (1 dp)

$\frac{dV}{dt} = \frac{dV}{dr} \times \frac{dr}{dt} \quad \frac{dV}{dr} = 4\pi r^2 =$ surface area, S

$\Rightarrow \frac{dV}{dt} = \frac{dr}{dt} \times S$, i.e. proportional to S

(since $\frac{dr}{dt}$ constant)

Section 7

1 (a) If perpendicular, $\mathbf{u} \cdot \mathbf{v} = 0 \Rightarrow 10 - 4t - 3s = 0$
i.e. $4t + 3s = 10$

(b) If parallel, $\frac{5}{2} = \frac{-4}{t} = \frac{s}{-3} \Rightarrow t = \frac{-8}{5}$, $s = \frac{-15}{2}$

2 (a) Subtracting, direction vector is $4\mathbf{j} - 3\mathbf{k}$
$\mathbf{r} = (\mathbf{i} + 2\mathbf{j} + 3\mathbf{k}) + \lambda(4\mathbf{j} - 3\mathbf{k})$

(b) $\cos\theta = \frac{(\mathbf{i} - 2\mathbf{j} + 2\mathbf{k}) \cdot (4\mathbf{j} - 3\mathbf{k})}{\sqrt{1+4+4}\sqrt{16+9}} = \frac{-14}{15}$

Acute angle, $\theta = 21°$ (nearest)

3 x-coords: $1 + s = -2 - 3t$ ... ①
y-coords: $2 - s = 1 + t$ ... ②
Adding $\quad 3 = -1 - 2t$
$\Rightarrow t = -2$

and point is $-2 - 3(-2) = 4$ and $1 - 2 = -1$
i.e. position vector is $4\mathbf{i} - \mathbf{j}$

Angle given by $\cos\theta = \frac{(\mathbf{i} - \mathbf{j}) \cdot (-3\mathbf{i} + \mathbf{j})}{\sqrt{1^2 + (-1)^2}\sqrt{(-3)^2 + 1^2}}$

$= \frac{-3-1}{\sqrt{2}\sqrt{10}} = \frac{-2}{\sqrt{5}}$

Ignoring negative sign for acute angle, $\theta = 26.6°$

4 (a) If a rectangle, OL is perpendicular to LM
$\overrightarrow{LM} = (5\mathbf{i} + \mathbf{j} + c\mathbf{k}) - (2\mathbf{i} - 3\mathbf{j} + 3\mathbf{k})$
$= 3\mathbf{i} + 4\mathbf{j} + (c-3)\mathbf{k}$
$\overrightarrow{OL} \cdot \overrightarrow{LM} = (2\mathbf{i} - 3\mathbf{j} + 3\mathbf{k}) \cdot (3\mathbf{i} + 4\mathbf{j} + (c-3)\mathbf{k})$
$= 6 - 12 + 3(c - 3) = 0 \Rightarrow c = 5$

(b) $\overrightarrow{ON} = \overrightarrow{LM}$ (since rectangle)
$= 3\mathbf{i} + 4\mathbf{j} + 2\mathbf{k}$, since $c = 5$

(c) Direction $\overrightarrow{MN}$ is $-2\mathbf{i} + 3\mathbf{j} - 3\mathbf{k}$
$\Rightarrow$ line equation is
$\mathbf{r} = (3\mathbf{i} + 4\mathbf{j} + 2\mathbf{k}) + t(-2\mathbf{i} + 3\mathbf{j} - 3\mathbf{k})$

5 (a) Direction $\overrightarrow{AC}$ is $(\mathbf{i} + 2\mathbf{j} + \mathbf{k}) - (5\mathbf{i} + \mathbf{j} + 2\mathbf{k})$
$= -4\mathbf{i} + \mathbf{j} - \mathbf{k}$
Direction $\overrightarrow{BC}$ is $(\mathbf{i} + 2\mathbf{j} + \mathbf{k}) - (-\mathbf{i} + 7\mathbf{j} + 8\mathbf{k})$
$= 2\mathbf{i} - 5\mathbf{j} - 7\mathbf{k}$
$\Rightarrow$ equations of lines are
l_1: $\mathbf{r} = \mathbf{i} + 2\mathbf{j} + \mathbf{k} + s(-4\mathbf{i} + \mathbf{j} - \mathbf{k})$
l_2: $\mathbf{r} = \mathbf{i} + 2\mathbf{j} + \mathbf{k} + t(2\mathbf{i} - 5\mathbf{j} - 7\mathbf{k})$

(b) $\cos\theta = \frac{(-4\mathbf{i} + \mathbf{j} - \mathbf{k}) \cdot (2\mathbf{i} - 5\mathbf{j} - 7\mathbf{k})}{\sqrt{16+1+1}\sqrt{4+25+49}} = \frac{-8-5+7}{\sqrt{18}\sqrt{78}}$

$= \frac{-6}{\sqrt{18}\sqrt{78}}$

ignoring the minus sign for an acute angle, $\theta = 81°$ (nearest)

6 Direction ratio is $(4 - 1)\mathbf{i} + -6\mathbf{j} + (-8 - 1)\mathbf{k}$
$= 3\mathbf{i} - 6\mathbf{j} - 9\mathbf{k}$
$= \mathbf{i} - 2\mathbf{j} - 3\mathbf{k}$

Equation is
$$\mathbf{r} = (\mathbf{i} + 6\mathbf{j} + \mathbf{k}) + t(\mathbf{i} - 2\mathbf{j} - 3\mathbf{k})$$

If intersect, x: $s = 1 + t$
y: $2s = 6 - 2t$
z: $-s = 1 - 3t$

$\Rightarrow s = 2, t = 1$ satisfies all three.

Intersect at $(2\mathbf{i} + 4\mathbf{j} - 2\mathbf{k})$

Angle is given by $\cos\theta = \dfrac{(\mathbf{i} + 2\mathbf{j} - \mathbf{k}) \cdot (5\mathbf{i} + a\mathbf{j} + 5\mathbf{k})}{\sqrt{1+4+1}\sqrt{25+a^2+25}}$

$= \dfrac{5 + 2a - 5}{\sqrt{6}\sqrt{50+a^2}} = \dfrac{2a}{\sqrt{6}\sqrt{50+a^2}}$

If $\theta = 60°$, $\cos\theta = \dfrac{1}{2} \Rightarrow \dfrac{2a}{\sqrt{6}\sqrt{50+a^2}} = \dfrac{1}{2} \Rightarrow$

$4a = \sqrt{6}\sqrt{50+a^2}$

Squaring both sides, $16a^2 = 6(50 + a^2)$

$10a^2 = 300 \Rightarrow a = \sqrt{30} \; (>0)$

7 (a) Direction ratio $\overrightarrow{AB}$ is $(6, 6, 12) \equiv (1, 1, 2)$

$\Rightarrow$ an equation is $\mathbf{r} = (24\mathbf{i} + 6\mathbf{j}) + \lambda(\mathbf{i} + \mathbf{j} + 2\mathbf{k})$

(b) If P is on line, coords are $(24 + t, 6 + t, 2t)$ for some t

$\Rightarrow \overrightarrow{CP} = (6 + t)\mathbf{i} + t\mathbf{j} + (2t - 36)\mathbf{k}$

(c) $\overrightarrow{CP} \cdot \overrightarrow{AB} = 0 \Rightarrow (6 + t) + t + 2(2t - 36) = 0$
$6t = 66 \Rightarrow t = 11$

Coords P are $(35, 17, 22)$

(d)

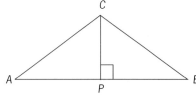

Area Δ is $\dfrac{1}{2}$ base $\times$ height $= \dfrac{1}{2} \times |\overrightarrow{AB}| \times |\overrightarrow{CP}|$

$|\overrightarrow{AB}| = \sqrt{6^2 + 6^2 + 12^2} = 6\sqrt{6}$

$|\overrightarrow{CP}| = \sqrt{17^2 + 11^2 + 14^2} = \sqrt{606}$

$\Rightarrow$ Area is $3\sqrt{6} \times \sqrt{606} = 181$ (3 s.f.)

8 (a) $\overrightarrow{OA} = \mathbf{i} + 2\mathbf{j} + 3\mathbf{k}$, so angle between $\overrightarrow{OA}$ and line given by

$\cos\theta = \dfrac{(\mathbf{i} + 2\mathbf{j} + 3\mathbf{k}) \cdot (-\mathbf{i} + \mathbf{j} + \mathbf{k})}{\sqrt{14}\sqrt{3}} = \dfrac{4}{\sqrt{42}}$

$\Rightarrow \theta = 51.9°$

(b) If P is on line, position vector is
$(1 - P)\mathbf{i} + (2 + P)\mathbf{j} + (3 + P)\mathbf{k}$

If perpendicular, $[(1 - P)\mathbf{i} + (2 + P)\mathbf{j} + (3 + P)\mathbf{k}] \cdot [-\mathbf{i} + \mathbf{j} + \mathbf{k}] = 0$

$\Rightarrow P - 1 + 2 + P + 3 + P = 0 \Rightarrow P = \dfrac{-4}{3}$

$\Rightarrow \overrightarrow{OP} = \dfrac{7}{3}\mathbf{i} + \dfrac{2}{3}\mathbf{j} + \dfrac{5}{3}\mathbf{k}$

(c) If Q is on line, position vector is $(1 - Q)\mathbf{i} + (2 + Q)\mathbf{j} + (3 + Q)\mathbf{k}$.

Length is $5 \Rightarrow (1 - Q)^2 + (2 + Q)^2 + (3 + Q)^2 = 25$

$1 - 2Q + Q^2 + 4 + 4Q + Q^2 + 9 + 6Q + Q^2 = 25 \Rightarrow 3Q^2 + 8Q - 11 = 0$

$(3Q + 11)(Q - 1) = 0 \Rightarrow Q = 1$ or $-\dfrac{11}{3}$

$\overrightarrow{OQ} = 3\mathbf{j} + 4\mathbf{k}$ or $\dfrac{14}{3}\mathbf{i} - \dfrac{5}{3}\mathbf{j} - \dfrac{2}{3}\mathbf{k}$

(d) $\overrightarrow{OR} = (1 - \lambda)\mathbf{i} + (2 + \lambda)\mathbf{j} + (3 + \lambda)\mathbf{k}$

$\overrightarrow{OS} = (1 - 2\lambda)\mathbf{i} + (2 + 2\lambda)\mathbf{j} + (3 + 2\lambda)\mathbf{k}$

If perpendicular,

$(1 - \lambda)(1 - 2\lambda) + (2 + \lambda)(2 + 2\lambda) + (3 + \lambda)(3 + 2\lambda) = 0$

$\Rightarrow 1 - 3\lambda + 2\lambda^2 + 4 + 6\lambda + 2\lambda^2 + 9 + 9\lambda + 2\lambda^2 = 0$

$6\lambda^2 + 12\lambda + 14 = 0$

Discriminant, $12^2 - 4 \times 6 \times 14 < 0 \Rightarrow$ no solution.

9 (a) $[\mathbf{r} = (2\mathbf{i} + \mathbf{j} + \mathbf{k}) + t(-2\mathbf{i} + 4\mathbf{j} + 2\mathbf{k})]$

i.e. $\mathbf{r} = (2\mathbf{i} + \mathbf{j} + \mathbf{k}) + t(\mathbf{i} - 2\mathbf{j} - \mathbf{k})$ is one possibility

(b) For x coordinate to be 4, t in last equation is 2. This gives position vector of $(4\mathbf{i} - 3\mathbf{j} - \mathbf{k})$, i.e. S is on line.

(c) $\overrightarrow{PQ} = -2\mathbf{i} + 4\mathbf{j} + 2\mathbf{k}$

$\overrightarrow{RS} = -\mathbf{i} + \mathbf{j} - 3\mathbf{k}$

$\overrightarrow{PQ} = 2 + 4 - 6 = 0$ $\therefore$ perpendicular

(d) $\overrightarrow{PQ} = -2\mathbf{i} + 4\mathbf{j} + 2\mathbf{k}$

$\overrightarrow{RQ} = -5\mathbf{i} + 9\mathbf{j} + \mathbf{k}$ (note both into Q)

Angle between is given by $\cos\theta =$

$\dfrac{(-2\mathbf{i} + 4\mathbf{j} + 2\mathbf{k}) \cdot (-5\mathbf{i} + 9\mathbf{j} + \mathbf{k})}{\sqrt{24}\sqrt{107}}$

$= \dfrac{48}{\sqrt{24}\sqrt{107}}$

$\Rightarrow \theta = 18.7°$

10 (a) $\lambda = 2$ for x-coordinate

This gives $7\mathbf{i} - \mathbf{j} + 2\mathbf{k}$, i.e. A is on line

(b) $AB^2 = (10 - 7)^2 + (-2 + 1)^2 + (2 - 3)^2 = 11$

$\Rightarrow AB = \sqrt{11}$

(c) $\overrightarrow{AB} = 3\mathbf{i} - \mathbf{j} - \mathbf{k}$

$\Rightarrow$ angle given by $\cos\theta = \dfrac{(3\mathbf{i} - \mathbf{j} - \mathbf{k}) \cdot (3\mathbf{i} - \mathbf{j} + \mathbf{k})}{\sqrt{11}\sqrt{11}}$

$= \dfrac{9}{11} \Rightarrow \theta = 35.1°$

(d)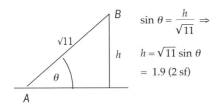

$\sin \theta = \dfrac{h}{\sqrt{11}} \Rightarrow$

$h = \sqrt{11} \sin \theta$

$= 1.9$ (2 sf)

11 (a) If perpendicular, $\lambda(1-\lambda) + (2\lambda-1) 3\lambda - (4\lambda-1) = 0$

$-\lambda^2 + \lambda + 6\lambda^2 - 3\lambda - 4\lambda + 1 = 0 \Rightarrow 5\lambda^2 - 6\lambda + 1 = 0$

$(5\lambda-1)(\lambda-1) = 0 \Rightarrow \lambda = 1$ or $\lambda = \dfrac{1}{5}$

(b) $\vec{OA} = 2\mathbf{i} + 3\mathbf{j} - \mathbf{k},\ \vec{OB} = -\mathbf{i} + 6\mathbf{j} + 7\mathbf{k}$

$\Rightarrow \vec{AB} = -3\mathbf{i} + 3\mathbf{j} + 8\mathbf{k}$

(c)

Want $-\vec{OA}$ for same direction.

$\cos \theta = \dfrac{(-3\mathbf{i} + 3\mathbf{j} + 8\mathbf{k}) \cdot (-2\mathbf{i} - 3\mathbf{j} + \mathbf{k})}{\sqrt{82}\,\sqrt{14}} = \dfrac{5}{\sqrt{82}\,\sqrt{14}}$

$\Rightarrow \theta = 82°$ (nearest)

12 (a) If intersect, y-coordinates must be the same,

$5 + \lambda = 1 \Rightarrow \lambda = -4$, x-coordinates same,

$\Rightarrow 12 + 2\lambda = 1 + 3\mu \Rightarrow \mu = 1$, giving position vector of A as $4\mathbf{i} + \mathbf{j} - 3\mathbf{k}$

(b) $\cos \theta = \dfrac{(2\mathbf{i} + \mathbf{j} + \mathbf{k}) \cdot (3\mathbf{i} - \mathbf{k})}{\sqrt{6}\,\sqrt{10}} = \dfrac{5}{\sqrt{60}} \Rightarrow \theta = 50°$

(nearest)

(c)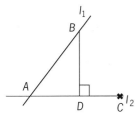

If on l_2, $\vec{OD} = (1+3d)\mathbf{i} + \mathbf{j} + (-2-d)\mathbf{k}$

$\Rightarrow \vec{BD} = (-15+3d)\mathbf{i} - 6\mathbf{j} + (-5-d)\mathbf{k}$

perpendicular to l_2,

$(-15+3d) \times 3 + (-5-d) \times (-1) = 0$

$-45 + 9d + 5 + d = 0 \Rightarrow d = 4$

$\Rightarrow \vec{OD} = 13\mathbf{i} + \mathbf{j} - 6\mathbf{k}$

(d) On line l_2, A has parameter $\mu = 1$, D $\mu = 4$, C $\mu = 7$. D is mid-point of $AC \Rightarrow \triangle ABC$ isosceles.

13 (a) $\mathbf{r} = (2\mathbf{i} + \mathbf{j} + 4\mathbf{k}) + \lambda(\mathbf{i} - \mathbf{j} - 2\mathbf{k})$

(b) $\dfrac{x-7}{-1} = \dfrac{y+6}{2} = \dfrac{z+4}{1}$

$\Rightarrow \mathbf{r} = (7\mathbf{i} - 6\mathbf{j} - 4\mathbf{k}) + \mu(-\mathbf{i} + 2\mathbf{j} + \mathbf{k})$

(c) $x:\ 2 + \lambda = 7 - \mu$

$y:\ 1 - \lambda = -6 + 2\mu$

Adding $3 = 1 + \mu \Rightarrow \mu = 2,\ \lambda = 3$

This gives z-coordinate of -2 and -2, so intersect. Position vector of point of intersection is $5\mathbf{i} - 2\mathbf{j} - 2\mathbf{k}$.

(d) If on l, coordinates are $(2+d, 1-d, 4-2d)$

$\Rightarrow (2+d)^2 + (1-d)^2 + (4-2d)^2 = (\sqrt{17})^2$

$4 + 4d + d^2 + 1 - 2d + d^2 + 16 - 16d + 4d^2 = 17$

$6d^2 - 14d + 4 = 0$

$3d^2 - 7d + 2 = 0$

$(3d - 1)(d - 2) = 0 \qquad d = \dfrac{1}{3}$ or $d = 2$

i.e. P and Q have position vectors, $\dfrac{7}{3}\mathbf{i} + \dfrac{2}{3}\mathbf{j} + \dfrac{10}{3}\mathbf{k}$ and $4\mathbf{i} - \mathbf{j}$

14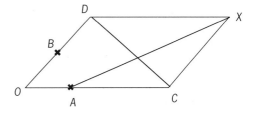

$\vec{AX} = \vec{AC} + \vec{CX}$

$= 2\vec{OA} + \vec{OD} = 2\vec{OA} + 2\vec{OB}$

$= 8\mathbf{i} + 8\mathbf{j} + 8\mathbf{k} = 8(\mathbf{i} + \mathbf{j} + \mathbf{k})$

$\Rightarrow$ parallel to $\mathbf{i} + \mathbf{j} + \mathbf{k}$

$\vec{AX}:\ \mathbf{r} = (3\mathbf{i} + 2\mathbf{j} + \mathbf{k}) + s(\mathbf{i} + \mathbf{j} + \mathbf{k})$

$\vec{CD}:\ \vec{OC} = 3\vec{OA} = 9\mathbf{i} + 6\mathbf{j} + 3\mathbf{k}$

$\vec{OD} = 2\vec{OB} = 2\mathbf{i} + 4\mathbf{j} + 6\mathbf{k}$

$\Rightarrow \vec{CD}:\ \mathbf{r} = (9\mathbf{i} + 6\mathbf{j} + 3\mathbf{k}) + t(7\mathbf{i} + 2\mathbf{j} - 3\mathbf{k})$

They lie in a plane, so in two dimensions must intersect.

Simultaneously, $3 + s = 9 + 7t$

$2 + s = 6 + 2t$

$\Rightarrow 1 = 3 + 5t,\ t = \dfrac{-2}{5},\ s = \dfrac{16}{5}$

Position vector of point of intersection is

$\dfrac{31}{5}\mathbf{i} + \dfrac{26}{5}\mathbf{j} + \dfrac{21}{5}\mathbf{k}$

$\vec{AB} = -2\mathbf{i} + 2\mathbf{k} \Rightarrow \cos \theta = \dfrac{(-\mathbf{i} + \mathbf{k}) \cdot (\mathbf{i} + \mathbf{j} + \mathbf{k})}{\sqrt{2}\,\sqrt{3}} = \dfrac{0}{\sqrt{6}}$

$\Rightarrow \theta = 90°$

15 (a) Putting $\mu = -2$ gives $4\mathbf{i} - \mathbf{j} + \mathbf{k}$

(b) Rewriting l_2 as $\mathbf{r} = -9\mathbf{j} + 13\mathbf{k} + v(\mathbf{i} + 2\mathbf{j} - 3\mathbf{k})$

also gives $4\mathbf{i} - \mathbf{j} + \mathbf{k}$ when $v = 4$

(c) λ: $v = -3 + 4\lambda$ $\Rightarrow -9 = 1 - 5\lambda, \lambda = 2,$
y: $-9 + 2v = -5 + 3\lambda$ $v = 5$

z-coordinate for l_2 is $13 - 3v = 13 - 15 = -2$

l_3 is $-4 + \lambda = -4 + 2 = -2$

i.e. intersect. Position vector of point of intersection B is

$5\mathbf{i} + \mathbf{j} - 2\mathbf{k}$

(d) $AC^2 = 3^2 + 1^2 + 4^2 = 26$
$BC^2 = 4^2 + 3^2 + 1^2 = 26$ $\Rightarrow AC = BC$

(e)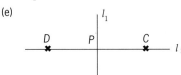
$\overrightarrow{AC} = -3\mathbf{i} - \mathbf{j} - 4\mathbf{k}$
$\overrightarrow{BC} = -4\mathbf{i} - 3\mathbf{j} - \mathbf{k}$

$\Rightarrow \cos\theta = \dfrac{12 + 3 + 4}{26} \Rightarrow \theta = 43°$ (nearest)

(f) D is mid-point of AB, i.e. $\left(\dfrac{9}{2}, 0, \dfrac{-1}{2}\right)$

16 (a) $\mathbf{r} = (5\mathbf{i} + 3\mathbf{j}) + \lambda(7\mathbf{i} + 7\mathbf{j} - 7\mathbf{k})$
$= (5\mathbf{i} + 3\mathbf{j}) + \lambda(\mathbf{i} + \mathbf{j} - \mathbf{k})$

(b) Putting directions, $(\mathbf{i} + \mathbf{j} - \mathbf{k}) \cdot (\mathbf{i} + 2\mathbf{j} + 3\mathbf{k}) = 0$

$\therefore$ perpendicular

(c) x: $5 + \lambda = 1 + \mu$
y: $3 + \lambda = -3 + 2\mu$ $\Rightarrow 2 = 4 - \mu, \mu = 2, \lambda = -2$
z: l_1 $-\lambda = 2$: l_2 $-4 + 3\mu = 2$

i.e. intersect, position vector $3\mathbf{i} + \mathbf{j} + 2\mathbf{k}$, call this P.

(d) $\mu = 1$ for x-coordinate. This gives $2\mathbf{i} - \mathbf{j} - \mathbf{k}$ as required

(e)

On l_2, C is $\mu = 1$, P is $\mu = 2$

so D is $\mu = 3$, i.e. $4\mathbf{i} + 3\mathbf{j} + 5\mathbf{k}$